T0189610

YORUBA IN DIASPORA

CONTEMPORARY ANTHROPOLOGY OF RELIGION

*A series published with the Society for the
Anthropology of Religion*

Robert Hefner, Series Editor
Boston University
Published by Palgrave Macmillan

Body / Meaning / Healing
By Thomas J. Csordas

*The Weight of the Past: Living with History in
Mahajanga, Madagascar*
By Michael Lambek

*After the Rescue: Jewish Identity and Community in
Contemporary Denmark*
By Andrew Buckser

Empowering the Past, Confronting the Future
By Andrew Strathern and Pamela J. Stewart

Islam Obscured: The Rhetoric of Anthropological Representation
By Daniel Martin Varisco

*Islam, Memory, and Morality in Yemen: Ruling
Families in Transition*
By Gabrielle Vom Bruck

*A Peaceful Jihad: Negotiating Identity and Modernity
in Muslim Java*
By Ronald Lukens-Bull

The Road to Clarity: Seventh-Day Adventism in Madagascar
By Eva Keller

Yoruba in Diaspora: An African Church in London
By Hermione Harris

An African Church in London

Hermione Harris

First published in 2006 by
PALGRAVE MACMILLAN™
175 Fifth Avenue, New York, N.Y. 10010 and
Houndmills, Basingstoke, Hampshire, England RG21 6XS
Companies and representatives throughout the world.

PALGRAVE MACMILLAN is the global academic imprint of the Palgrave Macmillan division of St. Martin's Press, LLC and of Palgrave Macmillan Ltd. Macmillan® is a registered trademark in the United States, United Kingdom and other countries. Palgrave is a registered trademark in the European Union and other countries.

ISBN 978-1-349-53550-7   ISBN 978-0-230-60104-8 (eBook)
DOI 10.1057/9780230601048

Library of Congress Cataloging-in-Publication Data
Harris, Hermione.
    Yoruba in diaspora: an African church in London/Hermione Harris.
        p. cm.—(Contemporary anthropology of religion)
    Includes bibliographical references and index.

        1. Yoruba (African people)—England—London—Religion. 2. London (England)—Religion. I. Title. II. Series.
BL980.G7H37 2006
289.9'3089963330421—dc22   2006041563

A catalogue record for this book is available from the British Library.

Design by Newgen Imaging Systems (P) Ltd., Chennai, India.

First edition: September 2006

10 9 8 7 6 5 4 3 2 1

For Alexis and Anna, and in memory of Marc

This page intentionally left blank

Contents

Acknowledgments

My principal thanks go to Baba Aladura S.A. Abidoye and to all the elders and members of the Cherubim and Seraphim, past and present, for welcoming me into the church. So many members spent hours of patient explanation, and I would not have survived the rigors of fieldwork without the care and humor of those who became good friends. I hope they will find something of use in the result of their generosity.

I am grateful to Professor John Middleton for comments on an early draft, and to Professor Richard Fardon for its appraisal when I returned to my material after several years. I then had an invaluable introduction to a changed anthropology through discussions of drafts with Andrea Cornwall, Joss Gamble, Kevin Latham, Jeff Roberts, and members of the SOAS research seminar. Thank you to Akin Oyetade for helpful and stimulating sessions on Yoruba language and culture. Juliet Ash, Annette Barnes, André Brink, Hugh Brody, Karen Brudney, Roswith Gerloff, Fiona Macintosh, Abiola Ogunsola, and Ellen Ross also offered encouragement, references, and useful comments on various chapters. Heartfelt thanks to Kate Hardcastle for deciphering and typing the bibliography and more, and to Lee Corless for dealing with computer crises.

The main "without whom" is Professor J.D.Y. Peel, long-time mentor and friend. Sharing his unrivalled knowledge of Yoruba political and religious history, he has provided constant intellectual and moral support. I thank him for his boundless patience with this long project.

My parents, Elizabeth and the late William Harris, whose generosity has given me the space and time to write, have lived with this book at one remove for many years. Jean McCrindle, at closer quarters, helped me keep going in numerous ways. I also thank Anna and Alexis Karlin for their encouragement, and for putting up with an often distracted mother. Marc Karlin did not live to celebrate publication, but never lost faith that the day would come. Marc—thank you.

This page intentionally left blank

Chapter 1

Introduction

And to our Sister Harris. I saw a tall ladder stretching up, and you were halfway up the ladder. You were saying that this is too big, it is too tall, and I want to come down. But the Lord says that you must pray to be able to go forward, and that you will finish your work with success.

Surely even Elder Ilelaboye, interpreting the trance-inspired images of his vision as a message from God, could not have imagined just how long the ladder would be. He was speaking at the end of 1969, a few months into my fieldwork on the Cherubim and Seraphim, a Nigerian church in London. I held onto the rungs until 1974, by which time several chapters of what was then a Ph.D. thesis were already written—but then I jumped off. It was not until the 1990s that I clambered on again, and found that I had to go back to the bottom step to rethink, rewrite, and re-research before I could finally reach the top.

There were various reasons for my starting the ascent once more. Looking at the sagging shelves of notes and documents, tapes and transcripts, the choice of their destination was between my newly acquired computer—or the bin. But too much work had already gone into the project (albeit nearly 20 years previously), not only by myself, but by scores of church members who had spent time they did not have to help me understand their world. Struggling to gain the professional qualifications for which they had come to Britain, the majority was also working to support themselves and their families. Their ladders, too, were long. The commitment I had made to record something of the lives of Yoruba worker-students and the history and practice of their church could not be abandoned lightly.

In the intervening decades, there has been a surge of publications on the Yoruba, the main ethnic group in southwestern Nigeria, and one of the largest linguistic groups in Africa as a whole.[1] I therefore imagined that my material might have been made redundant by subsequent literature on Yoruba in Britain, given the considerable growth of the Nigerian community, which now includes professionals and entrepreneurs as well as students. But it is striking, both at the time of my original research and in the intervening years, how little has been published on Yoruba communities in London. In Carey's *Colonial Students* (1956), Craven's *West Africans in London* (1968), and Mayo's article on "West African Voluntary Associations" (1969), Yoruba are subsumed in the general category of West Africans, as they are in Killingray's later chapter on "Africans in the United Kingdom" (1994b). In the plethora of analyses of race and ethnicity, now as then, the Nigerian diaspora is hardly mentioned. This silence is partly because the West African community has in the past largely been a transitory, not a settler, population, which was, as Smith commented, "very small, and therefore extremely difficult to survey" (1977: 21). The models of social adaptation and integration used by Carey and Craven were inappropriate for a community not intending to settle. But even in the following fashion of ethnic pluralism and the study of minority communities during the latter part of the twentieth century, Yoruba seem to have exercised little fascination. Many of the present generation of Yoruba in London contemplate permanent settlement, but still as yet have not attracted research.

One explanation for this neglect seems to be that Yoruba have not, as a group, constituted a problem either as a topic for cultural studies or for social policy. One of the mainsprings of research on race has been the threat of black unrest: the 1967 Political and Economic Planning (PEP) Report on racial discrimination anticipated that the second generation would not be "equally docile [as their parents] when faced with the frustrations and humiliations of discrimination; anger and violence, rather than self-effacement, may seem to them to be a more realistic response" (Daniel 1968: 14). Ten years later, the next PEP study warned of conditions that would lead to "conflict" and "open confrontation," "and lead [immigrants] to see themselves first and foremost as members of an oppressed class" (Smith 1977: 333). Yoruba were not incorporated in these studies. For reasons explored in chapter 2, they had, in general, excluded themselves from this "class." This tradition of political nonparticipation has largely continued; as an ethnic group, Yoruba have posed no public threat—nor have they captured the interest of radical researchers as providing potential militants.

But there are two specific areas defined as problematic by social agencies, which have produced some investigation. The first of these, from the 1950s to the early 1970s, was that of the student from overseas. In the pre- and immediate postwar period, the population of students from the then colonies was small and largely came from elite families. As the numbers grew, and the range of their class background broadened, so did associated problems, provoking worries over the educational and social experience of West Africans coming to Britain to study.[2] Many students took longer than they had hoped, and started families. This created the second problematic area: childcare. The Yoruba practice of fostering children with British families resulted in difficulties that confronted social services and the courts, generating research that still provides the principal ethnography on Yoruba in London.[3]

It therefore seemed essential to chart this period in Yoruba migration to Britain before this particular historical moment becomes erased by concern with current globalization. In chapter 9, I look at the contemporary diaspora of Yoruba who are leaving the collapsing Nigerian nation-state. But the body of the study focuses on the immediate post-Independence period, when young—and not-so-young—men and women saw education in Britain as their way into the new elite of Nigerian professions and state bureaucracy. The value of resuming a project after two decades is the comparisons it allows between these generations. My more recent research, undertaken from the end of the 1990s to the present, does not pretend to have the depth of the earlier fieldwork, but nevertheless provides both material and hypotheses unavailable elsewhere.

The same is true of my main topic: Yoruba Christianity in London. The Cherubim and Seraphim Church (C&S)[4] is one of the principal players in what have become known as AICs. The various referents of this acronym trace the development of the Cherubim and Seraphim, along with other Yoruba Aladura, or "praying" churches that originated in Western Nigeria.[5] First, in the 1960s and 1970s, the C&S was classed in the literature as an "African Independent Church," its members having moved out of the mainline Protestant mission churches from 1925 onward, to develop their own structure and mode of worship. Brought to London by students in 1965, the C&S was often later referred to as "Indigenous," stressing the ethnic character of congregations and practice rather than the early connection with orthodoxy. Toward the end of the last century, Aladura were described as "African Initiated" or "Instituted" Churches, hence leaving space for their aspirations to internal inclusivity as well as their external connection to the wider black church movement. Now, with the global spread of AICs, they may well come to be known by Ter Haar's term: "African International Churches" (1998: 24).

This development of the C&S and other Aladura over the last four decades or so means that they are now a ubiquitous part of London life. On Sundays, carloads of white-robed members are to be seen driving round the capital to the numerous branches of their churches. But this was not so when I first began my research. From my experience working and traveling in southern and eastern Africa in the 1960s, and recognizing African men and women on London streets, I knew that somewhere there must be an African religious organization. I set out to find it; it took me six months. Social workers, vicars, community organizers—all denied that any such thing existed, until a black Anglican priest recollected that some African group made use of a Congregational church in Shepherd's Bush. This was the C&S. I made contact with the secretary, asked if I might carry out research, went to a service—and found that, quite unprepared, I was under way.

At that time there were two main sources to provide me with Nigerian background: J.D.Y. Peel's classic study *Aladura* (1968) and H.W. Turner's two volumes on the history and theology of the Church of the Lord, another Aladura organization akin to C&S (1967). All these churches, together with various other West African groups, also refer to themselves as "spiritual" churches, from their emphasis on the Holy Spirit. This religious movement has been increasingly studied over the past three decades,[6] but all this research still focuses on West Africa, and neglects the Nigerian diaspora.

When I resumed my work in the 1990s, I imagined that it would have been preempted in the meantime. Two-third of black Christians in Britain are members of black-led churches.[7] But, despite their importance in their respective communities, they have still received little attention: a recent collection edited by Owusu, *Black British Culture and Society* (2000), does not even include religion in the index. Aladura organizations have also suffered from being subsumed in the general category of black churches, which are largely African-Caribbean. Gerloff (1992) and Kerridge (1995) include short pieces on Aladura in their surveys of black churches, as does Ter Haar in her study of African Christians in Europe (1998), but there is little that is specific. This elision of Yoruba and African-Caribbean experience leads to skewed analysis; Kalilombe, for example, argues in his "Black Christianity in Britain": "The origin and development of black Christianity in Britain is preponderantly the result of black people feeling alienated and marginalized" (1997: 317), a conclusion in line with earlier theories that accounted for black churches in terms of problems with assimilation or status deprivation. The C&S, with its norms of mutual assistance and elaborate, inclusive hierarchy, clearly provided support for an insecure, disadvantaged immigrant group. But Yoruba and African-Caribbean experience is not identical; Nigerian students in the 1970s were engaged in a class transition. C&S members saw themselves as a nascent national elite—as agents, not victims. But there is little to put the record straight: a doctoral thesis on the church in Birmingham,[8] two dissertations on the Church of the Lord,[9] all unpublished. My most recent research looked at the Nigerian Pentecostal Born-Again movement. There is a burgeoning literature on global Pentecostalism,[10] which includes the West African experience, but as yet little on Born-Again practice in London.

Another body of work developed over recent decades is that on new religious movements (NRMs).[11] But this concentrates on organizations, many with Eastern origins, which appeal to Western constituencies. African churches, springing from different roots and not attracting European members, do not fall easily into this category. Aladura and other spiritual churches have been labeled NRMs, but only in an African context.[12]

A significant development in the literature has been the growth of scholarship within the black churches themselves. As in other contexts, those with whom anthropologists have worked are now often equipped to comment on the results—or produce their own studies. There is a long tradition of Yoruba scholarship on indigenous society and religion,[13] and Aladura have always been interested in producing pamphlets on their history. Now the two have married: the work of J. Akinyele Omoyajowo (1978, 1982, 1984) on the Nigerian C&S is an example. In Britain, conferences attended by West African scholars who are also leaders of spiritual churches have produced invaluable exchanges.[14] Nevertheless, what is still lacking is a detailed ethnography of this major aspect of Yoruba life in London.

Spiritual Power

There also has been a lacuna in anthropological literature on the subject that forms the central focus of this study: spiritual power, the invisible energy with which the C&S God infuses his creation. Capable of being utilized for good or ill, power, it is said, may be solicited from gods or manipulated through speech, gesture, or material objects. In C&S thinking, evidence of power lies both in the assaults of unseen enemies that account for misfortune, and in the efficacy of prayer that protects and produces blessings. Its most dramatic manifestation is in entrancement, interpreted as possession by the power of the Holy Spirit.

Knowing nothing about either the Yoruba community or their religious practices when I first attended a C&S service in 1969, it was this aspect of the proceedings that immediately held my attention. Although no doubt succumbing to the tendency to exoticize the Other, criticized so fiercely in current anthropology, an uninitiated visitor could not avoid being arrested by evidence of the Spirit. From the gallery where I was sitting, I looked down on a sea of some 300 white-robed men and women, singing, dancing, clapping to the drums and organ, praying spontaneously in unison, or following the extemporary prayer from the leader of the service. As the atmosphere quickened, I noticed that a number of men and women began to tremble and shake; some were shouting unintelligibly, but others gasped out sentences, either in Yoruba or English.

The contents of these utterances, I saw, were being jotted down in a notebook by another member. Then, at a particular point in the service, first women and then men, in an orderly fashion, were asked to deliver their visions. Prefacing their delivery with "In the name of the Father, Son and Holy Spirit, as the prayer was in progress I saw . . . ," there followed a message to an individual or to the congregation, such as the one with which I opened this chapter. I was to learn that these revelations were based on images, "like cinema," which the visioners saw when in trance, and then interpreted according to Aladura conventions. They often enjoined unity, or right living, but also prescribed ritual to ensure well-being and success in the areas of health, fertility, exams, employment, housing, travel—an immediate window into the lifeworld of their recipients.

Even on this first occasion, I noticed that warnings of the assaults of enemies and the power of evildoers featured large. At one point in the service, a sister suddenly began to shriek, flinging her arms backward and forward, back bending, deep in trance:

Don't use them—don't use them—don't use what you don't
understand—black power, black power . . .

I was astonished. Even from my brief acquaintance with C&S members, it seemed that their personal ambitions and political orientation had little in common with the disciples of Stokely Carmichael or Malcolm X organizing the Black Power movement in Britain's cities at the time.[15] I was enlightened later by a member. Gossip had it that a certain elder was using occult literature,[16] invoking spirits of dubious provenance to supplement his own spiritual power. Although, to avoid conflict, he was not overtly named, this was a warning that he should not seek powers other than the Holy Spirit. The thrust of these accusations was partly moral: using sources of power other than holy water or oil, or personal prayer and fasting, smacked of indigenous "juju," Yoruba spells and incantations that Aladura condemned. But in addition, to summon forces not fully comprehended, without the necessary personal power to control the consequences, was to run the risk of spiritual or physical danger.

The following Sunday, these unseen perils were brought closer to home: summoned into the side-room after the service, I was told by a group of church elders (men in senior grades in the C&S hierarchy) that a sister had reported a dream concerning myself: pursued by a group of people, I had been captured, strung up by my hands on a telephone wire, and pushed to and fro. These aggressors, I was told, were evildoers, that is, witches or enemies posing as friends. The possible effects of their power were to be neutralized by prayer, duly said by an encircling group of elders.

From that moment, I became aware of the Cherubim and Seraphim's preoccupation with the pursuit of *agbara emi*, "spiritual power." Reworking my material some years later, it was still this concept of unseen energy that appeared to draw together different aspects of the C&S. The longer I looked at ritual practice, the more closely I read ritual texts, the more the salient features of this vitality emerged and the logic of its operation became apparent. To unpack an ineffable essence such as spiritual power is not inimical to the tropes of Yoruba religious discourse, in which "reification turns spiritual ideas to material things and makes them behave as [if] they are totally enfleshed" (Ilesanmi 1993: 65). Nevertheless, I am aware of the dangers of essentializing a fluid and ambiguous concept, of distorting C&S interpretations by overprivileging the notion of power, of presenting an idealized, intellectualist account of Aladura practice with few dissenting voices. It is certainly true that ordinary members would be unable to present the concept in such detail as I do in chapters 4 and 5. Nor would the adepts, the experienced elders on whom I heavily relied, spontaneously formulate the model in the way I have done. But neither have I, in the manner of some deconstructionist studies, imputed meaning to metaphor or significance to ritual acts where this was not explicitly confirmed by elders themselves. One of the advantages of undertaking fieldwork "at home" is the potential for checking emerging analysis with the community concerned. "You might not think it," said a prophet after one such session on *agbara*, "but that's actually what's going on." Elders did not always agree on detail—adepts have their own ritual styles—but the main conclusions were not disputed.

Reliance on indigenous experts has always been a key anthropological strategy, although in the early days the main informant may not have put in a personal appearance in the final account. Recent studies of Yoruba ritual in Nigeria such as that by Drewal (1992) are explicit about individual sources, and the relationship is more of colleagues or co-researchers than ethnographer and informant.[17] In the sensitive London context, those on whom I most relied preferred not to be forefronted, so do not appear as major characters. But their personalities and viewpoints molded my perceptions. For whether made apparent or not, fieldwork is a dialogue; the wresting of meaning from practice becomes a joint venture in which the informant is involved in self-realization as much as the ethnographer seeks to transcend his or her own cultural confines. Comments reached me from some of those with whom I talked: they found discussions "very interesting," discovering how much they knew about their church, and reflecting on this knowledge.

Whether because of their educational experience or personal curiosity, several of the elders displayed a high degree of reflexivity. For example, I did not encounter the problem that puzzled Carter (1997) in his study of the Celestial Church of Christ (a popular latecomer to the Aladura scene, much given to elaborate ritual): how could a church that clearly exhibited commonalities with traditional ritual so firmly reject charges of syncretism? The C&S reflected openly on the replacement of indigenous practice by Aladura. They made no apology that visions, especially those solicited from visioners with respect to particular problems, had supplanted Ifa, the Yoruba oracle; that Aladura ritual rivaled vernacular medicine and juju; that the power of the Christian God had now overtaken that of Yoruba deities, the *oriṣa*.

The forging of Christian forms from familiar cultural practice is effected in the C&S not by the reproduction of traditional ritual behavior, but through persistence of religious orientations and cosmological assumptions, those "dispositions which, integrating past experiences, function at every moment as a *matrix of perceptions, appreciations, and actions*" (Bourdieu 1977: 82–83). Central to these is the discourse of spiritual power. But this unseen energy is often as much a taken-for-granted matter for ethnographers as it is for many of those they have studied. In previous work on Aladura and similar West African churches, the significance of spiritual potency has been acknowledged: Beckman (1975: 114) writes that the theology of the Ghanaian spiritual church movement "can be summed up in a word: power." But it is not explored, or taken as a starting point for analysis. An exception is Hackett's study of the discourse of power among contemporary West African spiritual churches (1993). She demonstrates the centrality of the concept, the elision of ritual and secular power, and the various contexts in which it is represented. But she does not consider the dynamics of spiritual power per se—the principles by which it is assumed to operate.

Neither do the majority of studies of power in different cultural contexts take these into account. One early and perceptive cross-cultural study along these lines was attempted by Webster in his *Magic, a Sociological Study* (1948). But the analysis was trapped in the evolutionary framework of its time: from "magic" to "religion." Articles in Fogelson and Adams' *Anthropology of Power* (1977) lack a common theoretical perspective, and it is hard to tell whether differences discernable in the discourse of power proceed from the cultures concerned or from the categories employed in analysis. This collection shows that the notion of spiritual power is more salient in some societies than in others, but we do not know whether this is inherent in differing metaphysical structures or due to the extent of the reduction of indigenous culture by external forces. Writing over a decade later, Arens and Karp's *Creativity of Power* (1989) adopts a sharper focus on power as a cultural construct. Challenging the secular definitions assumed by Western social scientists, they argue that analysis of power should start with indigenous concepts: "power is how power means" (xv). But although they draw attention to the sources and exercise of ritual power, and note its universal transformative capacity, the articles in their volume do not enquire *how* this power is constructed or examine the particular shape it has in local imaginations.

Early studies of magic drew much on the concept of *mana*, a Polynesian term originally documented at the end of the nineteenth century, and then adopted as a generic term for concepts of innate, effective vitality in living beings, material objects, and invisible entities.[18] But Keesing, in "Rethinking Mana" (1984), argues that *mana* itself may not have been conceived as a "thing-like" potency, but more as a quality of efficacy, only realized through visible results. The oft-employed metaphor of electricity, he maintains, was not indigenous. Awareness of subtle differences in the concept of unseen energy warns against the derivation of metaconcepts from one particular cultural context for indiscriminate use in others. Fr. Placide Tempels, driven by a quest for effective evangelization in the 1940s, overemphasized the ubiquity of an all-pervasive *force vitale* in Africa rather than examining the dynamics of unseen vitality in its particular Congolese setting. His *Bantu Philosophy* (1969) had considerable influence over African scholars generalizing on the "African philosophy of life," which "conceives of the quality of 'force,' 'energy,' or 'power' " that "derives from God and runs through the whole of creation" (Echeknube 1987: 18). These conclusions are, however, contested in the debate within African scholarship over the extent to which this "life-force" is a universal feature of African epistemology.[19]

In Western Nigeria, however, the concept of power as a necessary vitality is omnipresent. For the C&S, *agbara emi mimo* "the power of the Holy Spirit" is a fundamental principle of general efficacy; the metaphor of electricity is their own. It forms the focus of ritual, the pivot of prayer, the subtext of all dreams and visions. It is also the basic dynamic behind the Aladura divinatory system of prophecy and prayer, whereby the visioner, empowered by the Spirit, reveals the activities of unseen forces. These may then be challenged or reinforced according to divine instruction through the enlisted power of God.

Although the nature of the unseen power remains culturally specific, I would argue that this circular oracular model, the dynamic relationship between etiology, oracle, and remedial action, is applicable to a wide range of African divination systems. Although rarely analyzed, the ontological ground is some notion of unseen energy articulating these aspects of the divinatory process both at an explanatory and performative level. However implicit, there will be some understanding of a spiritual force that influences both the diviner's operations, and the agents whose activities are so revealed, whether this vitality is personified or not.

Take the famous case of the Azande: in spite of the relative insignificance of deities, spirits, and ancestors in the Zande cosmology,[20] Evans-Pritchard records that it is the "play of mystic forces" that determines misfortune or success (1937: 340–341). These are revealed only through divination, where it is the "mystical potency"[21] of the poison administered to chickens, or the medicine in the entranced diviner's belly,[22] which operates the oracles. This dynamic *mbismo*, the "inherent power"[23] of persons and things,[24] lies behind all unseen agency;[25] it "bridges over the distance" both between witch and victim,[26] and between rectifying ritual and its result.[27] In another classic study, Turner's analysis of Ndembu divination (1975), malign agents are held responsible for misfortune through the "attribution of magical power to motivation";[28] it is the power of a spirit that shakes the diviner,[29] determining the fall of the objects in the divinatory basket— symbolic objects "invested with some animate quality"[30] whose "meanings rest ultimately on axiomatic beliefs in the existence of mystical beings and forces."[31] By way of introduction, Turner asserts that in "African thought . . . materiality is not inert but vital."[32] But the character of this vitality, which links the ideational with the performance, the ontology with the séance, remains opaque.

By emphasizing the dynamic quality of spiritual convictions, I am not suggesting that this is exclusively African. Much of Western New Age practice is essentially the pursuit of unseen powers with a modern gloss:

> Celestial Black Dragon Power, awaken its lightening bolt power, instantly create a power centre at home or work . . .
>
> The Field: The Quest for the Secret Force of the Universe, the key to life itself may lie in the vibrations that connect everything.
>
> (Programme for Mind Body Spirit Festival,
> London, 2002: 6)

> Modern physics affirms that everything in our world is formed by energy, and that it is the variations in its vibrational frequencies which make this energy able to manifest in all shades of density, ranging from rocks to gases. Everything then is an integral part of the Whole, constantly exchanging this basic common factor. By becoming aware of this process of exchange, we can employ it far more purposefully.
>
> (St. Aubin 1990: 21)[33]

But in the C&S this search is at once less self-conscious (as needing no justification) and implicitly more coherent (being grounded in indigenous epistemology). The pre-Christian Yoruba term for the power was *aṣe*; for C&S it is *agbara* that makes sense of prayer and prophecy. I would maintain that all oracular practice that goes beyond mere fortune-telling will be similarly informed by some notion of spiritual power as the effective principle behind causation and cure.

Returning to Anthropology

The connection between power and divination was apparent to me during my original research. But I had problems in deciding what to make of it. A relevant question might be not only why I started up the ladder again after so long, but why I slid off in the first place. The reasons were various, but in retrospect, I think that although I had so much already written, I could not find a way to deal with all of my material. I was stuck.

My theoretical aim was straightforward. I wanted to argue for religion as a significant factor in its own right, and to accord belief and ritual practice equal weight in the analysis of social action to political and economic factors. The way people interpret their reality, and therefore how they act upon it, has a dynamic, dialectical relationship with other aspects of their lifeworld, but is, I wished to show, an autonomous determinant both in the construction of social groups and the decisions of individuals. To relegate epistemology to a second level of analysis, it seemed to me, was to distort agency and distrust emic accounts of motivation. This is now a commonplace assumption, in need of no special pleading. But it was much less so in the early 1970s. On the one hand there was Marxism, and on the other, the functionalist heirs of Durkheim. Marxist perspectives were once again fashionable in academia, in forms ranging from the synthesis of structuralism and Marxism in French anthropology to the dependency theory of Latin American provenance. These currents stimulated new insights into the social relations of production, such as Gavin Williams's work on the political economy of Western Nigeria (1970, 1976). But what of religion? This was seen as ideology, relegated to a superstructural second order, which was held to mystify exploitation and inequality.[34] Although they stood at the opposite ends of the political spectrum, Marxists thus merged with structural functionalists in treating religion as essentially epiphenomenal, whether of political domination or social cohesion. Although many of the classic studies of religion in the early 1970s were more complex than a crude label implies, the underlying premise of the functionalism in which I was trained was reductionist. Ritual was often represented as "all purpose social glue," either binding communities together, or providing a cathartic release of conflict to enhance social solidarity.

Clearly, a sociology of religion must recognize what a religious practice or organization *does*. But it must also take into account what it *is* for its followers. Part of anthropological debate in the 1970s was over the definition of "religion." My own working model of the concept was substantive rather than functionalist: as beliefs and practices arising from the supposed operation of spiritual forces in the universe, whether invested in objects, words, or gestures, or embodied in spirits or gods. This emphasis bypassed the old debate on the distinction between "religion" and "magic,"[35] and replaced it with members' own interpretation of their ritual behavior. I was interested in how the *content* of ritual expression and the assumptions that lay behind action and speech affected the life course of individuals, as well as the development of a Yoruba organization in London. But current theory was addressing different questions.

When I finally returned to my research in the 1990s, I had been out of anthropology for nearly 20 years; that is, divorced from the academic discipline, for anthropological insights proved invaluable in the

Central American development projects with which I had been working. In hindsight, I can see that this practical application of anthropology reinforced my theoretical bent. The success of projects depended on the meaning assigned by a community both to their present economic, political, and social arrangements, and to proposed changes. Many a development scheme that has been based on theories of modernization but which has ignored local understandings has ended in failure. I then found that what had happened to anthropology in the interim helped me make sense of my data. The cultures that anthropologists had traditionally studied were changing, and old models no longer served. By the late 1970s, the much-mentioned theoretical shift from function to meaning was well under way. Although their theoretical approaches were distinct, the monographs on spirit possession with which I started—Lambek (1981), Danforth (1989), Boddy (1989), amongst others—shared this interpretive concern: to uncover the meaning behind social practice, what it signified not only to the community but to individuals. "Agency" put people and process in a dialectical relationship with systems and structures. The metaphor of text for approaching culture, popularized by Clifford Geertz (1973, 1983), embraced all social behavior; every aspect of life could be read for its symbolic relationship with others. In my own work it was the sermons, prayers, and visions that assumed particular new significance as multilayered texts, operating on both performative and metaphorical levels. Whereas their manifest content spoke of members' daily preoccupations, the imaginative universe they conveyed was one of the battle between good and evil forces, the nature of the power through which this conflict was conducted, and the principles upon which it was thought to operate. To start as I have done (chapters 4 and 5) with an outline of these assumptions, rather than with lived events, might seem to divide social practice into "worldview" and "behavior." But this epistemology is as essential a background for interpreting possession and ritual, as is the socioeconomic position of members in explaining the formation of the C&S.

The 1980s also saw a renewal of interest in African divination, which "continues to provide a trusted means of decision making, a basic source of vital knowledge" (Peek 1991a: 2). Typologies of oracles tend to set apart the possession of spirit mediums from mechanically operated oracles,[36] but it has also been recognized that the two methods act as structural equivalents.[37] In Aladura practice, consultation with visioners over the vicissitudes of life is part of the attraction of the C&S to its members, as well as to those who seek a divinatory service without identifying with the church. An appreciation of this oracular function of possession in African independent churches appeared from the 1960s,[38] but the process of divination through the Holy Spirit—how it operates in practice—has received surprisingly little attention.[39]

Earlier functionalist approaches emphasized the instrumentality of oracles;[40] Aladura prophecy, as all divination, certainly has both intended and unintended consequences in terms of personal and collective decision-taking and ordering of community life. But it also conveys Aladura epistemology, rendering personal problems meaningful by setting them into the C&S cosmological context. In recent studies, "attention is focused on the aspect of purposeful articulation of meaning" (Devisch 1985: 77), analyzing the *process* by which an oracle creates consensus from diversity, clarity from confusion, and arrives at a result. Instead of a prototypical presentation of divinatory procedure, contemporary ethnography explores examples of oracular discourse, showing how meaning is wrested from metaphor and ambiguity in the context of concrete cases.[41] To date, there have been no comparable studies of the divinatory aspect of spiritual churches. Through the pages of a prophet's diary, in chapter 8, I flesh out a model of the divination process with cases from his clientele.

Ethnography and the Self

The reliance on a record solicited from a practitioner in place of direct observation stems from the context of this study: for reasons of confidentiality, I could not attend or follow up individual consultations with visioners. But my fieldwork was also shaped by the academic climate of the times as well as the community concerned. Although I might have tussled with its constraints, I had begun my research within a particular disciplinary culture, which inevitably influenced what and how I observed, not only how I interpreted the findings. I was working, for example, before the women's movement challenged the andocentric bias in the collection of data. I had a position of responsibility in a female Band (a subgroup within the church), and formed close relationships with several women. But my main research relied on the male spiritual experts and elders of an organization that was led and controlled by men, rather than the more inaccessible experience of women.

Feminist theoreticians have also been in the forefront of the debate on the author in ethnographic writing, demanding a self-awareness of a situated researcher through the exercise of reflexivity: "the turning back on oneself, a process of self-reference" (Davies 1999: 4).[42] This freedom for the author is exemplified in rich and fascinating studies such as Cornwall's work with Yoruba women in Ado (1996) or Barber's experience of Yoruba popular theatre (2000), where the writer's presence is woven seamlessly into the account.

Other authors are more self-consciously autobiographical, such as Rabinow in his *Reflections on Fieldwork in Morocco* (1977), or adopt a confessional mode, as do the contributors to *Taboo: Sex, Identity and Erotic Subjectivity in Anthropological Fieldwork* (Kulik and Willson 1995). Then there is the "ego-ethnographic" genre, which charts the researcher's personal odyssey of self-discovery as in Stoller and Olkes's *In Sorcery's Shadow* (1987). One definition of hermeneutics—the explanatory, interpretive strategy of much contemporary anthropology—is that made by Ricoeur: "the comprehension of self by the detour of the comprehension of the other" (quoted in Rabinow 1977: 5); conversely: "the conscious use of the self as a resource for making sense of others" (Hervik 1994: 92).

Either way, in what follows I seem largely, and unfashionably, absent. Educated in an academic tradition where the first person was prohibited— "I think" was scored out by teachers in favor of "it could be argued that . . ."—I was trained in an anthropology that still saw itself as an objective science. "Personal" accounts of fieldwork tended to be relegated to journals, famously by Malinowski in *A Diary in the Strict Sense of the Term* (1967), novelized as in Bowen's *Return to Laughter* (1954), or regarded as academically inferior.[43] When working with the C&S, my fieldwork notebook labeled *Diary* soon became a record of others' experience, rather than my own. The most extreme case of this self-negation was my personal experience of possession, which took place during one late-night service in 1973. As it was one of the most "powerful" events of my life, I do not need a record to recall it, but my field notes enter members' questions and comments on the incident, but without a word on what happened to me, or my own response.

When writing the chapter on possession (6), I first included a personal account of my trance-experience. But the change of voice sat oddly, and I took it out. Nevertheless, it is there—as an understanding, not as an event. My own "sensory knowledge" of possession gave me an understanding of others' encounter with the Spirit, and also influenced my analysis. Whatever additional significance the possession phenomenon might have, for the C&S a primary meaning is that of *experienced* power. Experience, both physical and emotional, I also found, was now recognized as a significant aspect of ethnography, which helped me approach my material afresh.

In the "Field"

The gathering of this material was conditioned by the concrete conditions of the community with which I worked. Every "field" has its difficulties: I was faced with a scattered, unbounded community in a large city with no focus apart from intermittent meetings and services, and a web of personal relationships.

Nevertheless, becoming an active member of the church drew me immediately into a network of ritual, administrative, and social activity. For two years I had the experience of being completely immersed in a community that was part of my own society, but almost invisible to it. Through a member's landlord, I found a place to live in Stoke Newington: a divided room at the top of a large, dilapidated Victorian house, in which also lived one Yoruba and four African-Caribbean families in the sort of conditions experienced by many C&S: one barely functioning lavatory for all, no usable bath, but plenty of rot, crumbling plaster, and mice.

My most important piece of equipment was the telephone, a virtual equivalent of village street, market, or town bar where anthropologists hang out. Much of my data depended on reportage: secondhand accounts passed on over the phone; conversations, rather than just "being around." A pulsing web of communication across London between members conveyed the latest church news, summons to group prayers, announcements of meetings and gossip. The obligation to report a significant dream or vision either directly to the person featured, or to elders, in order for prayer to follow, provided opportunity for exchanging further information. Included in the network, news reached me fast. When once I was upset about a malicious rumor about myself that had come back to me, an elder consoled me: "If they didn't gossip about you, you wouldn't be one of them."

This particular accusation was that I had purchased my car from the proceeds of the Building Fund Committee (established to buy a prayer house) of which I was a member. What lay behind this and other rumors was distrust—a fear that as the only white member of the church I was a CID agent/Home Office spy. Suspicions were exacerbated by my tape-recording church services, a researcher's persistent questioning, and stories of a vision allegedly disclosing my true intent. These episodes revealed the extent of members' insecurity. Mistrust was legitimate; immigration rules were tightening, and some members may well have been sailing close to the wind.[44] The fear of official authority, "trouble," and deportation was ever-present. The C&S was also applying for registration as a religious body; elders were apprehensive that the church was under investigation as a covert political organization, which would threaten their official recognition. Many Yoruba have commented to me on the preoccupation with privacy amongst "our people"; personal enquiry, which would be the stuff of English polite conversation, can be considered impertinent or suspect. One C&S friend refused to take on life insurance as the form "asked too many questions." This initially made investigation into members' backgrounds problematic. When I asked the head of one male Band if members would answer four brief queries including date of arrival in England, the request was rejected as being "too personal." A premature attempt to administer a more detailed questionnaire to elders nearly ditched my fieldwork.

However, with time, and through the support of many members at both an official and personal level, these problems receded. The elders' survey (referred to as the elders' sample), which I conducted with 32 men, covered both personal background and church careers. Seventy-five members of two male Bands responded to a shorter written questionnaire handed out at a meeting. These two together formed the male members' sample—a total of 92 men, as there were elders amongst the members of the Bands. But even though not yielding quantitative data, far more information was gathered by constant discussion with both elders and "floor members" over two years, with more irregular contact for several more. Although I would have an agenda for each meeting, I found that relatively unstructured and openended discussion was more productive for deeper insights both on individual lives and C&S practice. Personal records—diaries, photo albums, correspondence—filled in individual stories. Although such confidences were vital to my understanding, to use directly the intimate knowledge I acquired of friends' personal histories and subterranean church events would be a betrayal of trust. I have attributed texts such as prayers and visions, together with comments on church affairs, to their speakers, but I have changed names when dealing with personal lives. Details of the C&S in Britain were filled in from church archives; the plethora of C&S pamphlets on the history and doctrine of the church brought from Nigeria (translated for me by elders where necessary) were supplemented by interviews with visiting leaders of Nigerian branches.

Over a period of two years I attended church occasions regularly: services, Sunday schools, Band meetings, committees, services in private houses, prayers organized for particular purposes. I became involved in church administration and social life. More intermittent participation continued for another four years. I also attended other Aladura churches[45] and interviewed their leaders. During 1999–2001, besides returning to the C&S, I attended services and events at four Pentecostal churches popular with Yoruba,[46] and discussed their organizations with pastors and members. During the first stage, I recorded and transcribed 26 services of different types. These included nearly 2,000 revelations. A sample of eight services of four kinds yielded 678 visions (referred to as the vision sample), which I analyzed in more detail. Sermons and Sunday schools were in English, but prayers and revelations were delivered in both Yoruba and English. To my shame I never learnt Yoruba. I had the unusual experience of starting fieldwork without knowing the ethnicity of the community concerned, and then found that everyone could speak English. So texts in the vernacular were translated for me by an elder. I have used these versions throughout, as they are closer to their English counterparts than a stricter rendering would be. When quoting verbatim from written or oral sources, I have retained original spellings and constructions, and dispensed with indicators of technical errors, such as *sic*. Amongst other sources, I have used Yoruba popular fiction[47] written in English, a genre that offers rich insight into contemporary life both in Nigeria and Britain, and the social and epistemological background to the Cherubim and Seraphim Church.

Chapter 2

"Stars of the World": Yoruba Worker-Students in Britain

When you come to England, you lock up your identity. In the work you do, and the way you live, you could be anybody. You must be prepared to take any job and sink to any depths—as long as you know what you are aiming at, and you don't let yourself lose sight of it.

(Elder Oguntulu)

Students and professional people! You are the Stars of the World!

(Sermon 1969)

When Yoruba students arrived in Britain in the first half of the 1960s, they envisaged a short, temporary stay in London before returning to Nigeria to join the ranks of the national elite. They came from a newly independent Nigeria with one aim in mind: to continue their education in the country by which they had been colonized for over half a century. What they were to experience was a Britain gradually slipping into economic decline, which afforded them little of the welcome they might have expected, and relegated them to the position of a black immigrant proletariat. As their stay became protracted, daily difficulties accumulated, frustrating their goal of gaining their qualifications and returning home. For members of the Cherubim and Seraphim it was their church that both helped them survive their problems and maintain a sense of their own professional identity and future.

Samuel Ajayi's Story

An elder in the C&S, Samuel Ajayi once told me of the turning point in his decision to come to England. It was in 1960, when he was working as a sanitary inspector in Lagos. One evening he arranged to meet an old school friend who had recently returned to Nigeria with a Law degree from London. His friend duly arrived in a handsome Peugeot to collect him after work, and both struggled to load Ajayi's battered bicycle into the boot. Ajayi was mortified. When his friend enthused about London, and his resultant success, Ajayi resolved to try it for himself. In 1972, he too was back in Lagos after 11 years in Britain, with qualifications as a mechanical engineer.

Ajayi was a serious man. Aware of his seniority, both in age and in the C&S hierarchy, he usually weighed his words carefully before he spoke. In services, while fellow elders danced and clapped, or were shaken by the Spirit, Teacher Ajayi would stand swaying slightly on the altar platform, small and solemn. The strains of the last ten years showed in his face, for, although he persevered until he achieved his aims, the experience had marked him.

Ajayi was born in 1930 in a village near Ibadan. He was already over 20 when he left school. His mother, a weaver and petty trader and the senior of four wives, was literate despite having had no formal education. His father had been converted and educated through the patronage of an Anglican pastor and had worked with a German trading company as well as having several stores of his own. The depression in the 1930s had hit both ventures, and interrupted Ajayi's education at a Church Missionary Society (CMS) school. But he eventually gained his West African School certificate and entry into the Lagos School of Hygiene. During the next seven years while working in Lagos, he studied for six O and three A levels in correspondence courses from England: "Overseas Tuition, Rapid Results, Wolsey Hall, all the lot." These were in arts subjects, as he was set on studying Law.

In 1962 Ajayi arrived in England. He had nothing arranged in advance, nor contact with any organization that could advise him. He had entered England as a student, having signed up with a secretarial college solely for the purpose. He found to his dismay that legal studies were difficult to obtain, as was a job, let alone one commensurate with his expectations. His savings were running out when he finally found work as a tea-boy/cleaner in a factory. After a wasted year, he decided to switch to engineering, and enrolled part time at South East London Technical College, whilst continuing work as a machine operator. He felt very isolated during this time. The C&S, which he had joined in Lagos in 1949, had not yet been established in London; although he knew other people from his region, who later were to join him as elders in the church, they did not visit him. At the time, he took this as a rebuff from people who were now better off than himself; it was only later that he understood their shame at their own straitened circumstances.

By 1964 he had saved enough to bring his wife over from Nigeria. On arrival, she found factory work, which enabled Ajayi to study fulltime for his Ordinary National Diploma (OND) in Mechanical Engineering. He was also taking four different correspondence courses, including Law and Company Secretaryship, which he continued in a desultory fashion throughout his stay in England, although he never sat for the exams. At this point he had little extra time to study, for in the vacations he would take a variety of low-paid jobs such as a fitter, hospital orderly or storekeeper, with temporary post office jobs at Christmas. He also became active in the C&S when it was founded in London in 1965.

He eventually achieved his OND, and continued working for his Higher National Certificate (HNC). But progress was slow: in 1968 when he failed an exam, he lost the grant that a sympathetic lecturer had helped him to obtain. So for the next two years he continued with a complicated combination of work, unemployment, and full- and parttime study. Each part of his life seemed to sabotage the other. As he was a student, he could not hope to get the kind of work that would give him the practical experience needed for a good post in Nigeria, except for the six months training he received as part of his course. His work not only took time and energy from his education, but at one time nearly cost him his residence in England when a less supportive lecturer discovered that he was contravening immigration rules by earning while still registered for fulltime study.

Financial worries were constant; apart from domestic demands, Ajayi was also sending money home for the education of his younger half-siblings. Accommodation was another problem: he had seven different addresses during his time in London. Generally he lived in rooms in multi-occupied houses with shared kitchens and bathrooms, moving on for reasons of expense, distance from work or college, poor conditions, or disputes with co-tenants or his West Indian and African landlords. Between 1965 and 1970, he managed to rent a small flat, but gave this up when his wife left him. His marriage, by the late 1960s, was going badly. His wife's family had been opposed to the match, and both he and Mrs. Ajayi were distressed that they had no children. In Mr. Ajayi's mind, these two facts were connected: it was the ill-will of her family, implemented through "invisible means" from Nigeria, that had stopped them from "having issue." A combination of these domestic worries, financial demands, news of his father's prolonged illness at home, and the intense pressure of exams made these years seem a long struggle. He was often exhausted: in college or workplace early in the morning; home by seven; study until three in the morning. His health suffered, he became depressed, and was terrified of having a breakdown: "I was already engulfed in the quicksands of agony. I was sinking down deep, deep. Sometimes I did not know where I was. I had no sleep. I just kept praying for the day that I will get out of this country, either dead or alive."

Mrs. Ajayi had her own problems. Having worked in catering to help support her husband, she had no qualifications of her own beyond some typing and dressmaking. Despite efforts of C&S members to mend the marriage, the relationship became untenable. Finally, in 1970, she left Ajayi to live with another man whose wife had remained behind in Nigeria. He treated her badly, but it was hard for her to return to Africa with neither a husband nor qualifications.

Meanwhile, church members set about trying to find Ajayi another wife, but the few single women in the church did not want to risk gossip implicating them in the separation. Working to save for his return to Nigeria, he had little opportunity to pursue a social life outside the church. So it was hard to meet other women, or to attract them when he did; his efforts with clubs and bureaux proved, he said, to be a waste of money.

In 1969, he had left college in order to start saving to go home by working as a laboratory technician at Hendon Technical College for £15 a week. This was not what he wanted, and he constantly saw less qualified white workers filling the posts for which he had applied: "All you get is good stories and then a refusal." He was still trying to complete his course with evening classes, but kept failing one particular subject, and so had to resit all the papers every time.

In 1970 he finally succeeded, passing with credits. He had already bought a trunk, and had started applying for jobs in Nigeria, for a vision by a C&S prophet had reassured him that he would soon be returning, and he believed that "the Holy One of Israel lieth not; he will never deceive me." Needing to earn more to prepare for this, he took a job at Ford's for £35 a week. This meant combining long hours of travel to Dagenham and shift-work on the assembly line with time at his desk. For he was still studying, aiming at his finals with the Institute of Mechanical Engineers, as well as embarking on a course with an institute of management to give him experience on the administrative side of engineering and assist his entry to his professional association in Nigeria.

By 1972 he was ready. Sailing from Liverpool, he was met on his arrival in Lagos by a delegation of C&S leaders and members of the church he had attended before he came to Britain. When Ajayi had left Nigeria, he had promised his family that he would only be gone for three years. In the event, he was away for over a decade, and never saw his father again. But his determination had finally been rewarded.

Despite his age, he was appointed to a senior post with the Department of Petroleum Resources at the Federal Ministry of Mines and Power. By the end of the 1970s he had remarried and had a child. Ajayi's story is not a composite case history, a combination of several accounts that recreates the "typical" subject as in older ethnographic style. But it well might have been. From the scores of life stories recounted to me by C&S members, the same patterns recur, the same problems emerge, which then reappear as the dilemmas that distressed students bring to the C&S. We will see them surfacing in members' private consultations with visioners and prophets, and as those circumstances that God, speaking through dream and vision, promises to ameliorate. They constitute the subtext of the "obstacles" in the way of progress, so often mentioned in prayer and revelation; they provide the context of the diffuse anxiety from which so many students suffer in pursuit of future success.

Aspirations, Education, and the Yoruba Elite

In the immediate post-Independence days when Ajayi left Nigeria, the average per capita income in the Western Region was £25 a year.[1] His starting salary at the Ministry was £2,000 per annum, an income level enjoyed by no more than 0.1 percent of Nigerians. In 1969–70, 19 prominent male church members returned home to jobs in regional or federal ministries, government and privately owned corporations, universities and training colleges. Only two had not found jobs by the end of 1970. Within 18 months of their return, half of them were earning as much or more than Ajayi. It was their time in Britain that had secured their entry into the new middle class of independent Nigeria.[2]

As technicians and professionals in corporations and state bureaucracies, it was the middle ranks of the Nigerian elite at which church members aimed. They did not aspire to the top ranks of Nigerian society, the most senior posts in ministries, state corporations, and the civil service, where salaries could reach £3,350 and more. This was the ruling elite. In theory, state employees could not engage in politics, but positions in the highest ranks of the medical, academic, and legal professions provided a stepping stone to a political career. Members of the C&S in London did not set their sights so high; none mentioned such political ambitions.

Nevertheless, despite a noticeable cultural sensitivity to ranking, the social space between these different strata of the elite appeared to be small; church members pointed out pictures in their albums of friends and kin in senior government positions, men and women they would encounter both at work and socially.[3] In discussions about their return to Nigeria, members identified more with these upper echelons than with their less educated kin. Certainly they aspired to a similar, even if more modest, lifestyle. Photographs sent back from returned church members showed neat bungalows on estates and campuses, or smartly furnished interiors of modern flats, rather than the communality of rural or small-town compound life, or rented rooms in crowded city neighborhoods. Built for colonial expatriates, these homes were only suitable for nuclear families. Although kin might lodge with an elite family in order to attend school, or to provide domestic help, such households contrast with the extended families found in most Yoruba dwellings. Those who intended to replace expatriates at Independence inherited not only their jobs, but also modeled their lifestyle on Western patterns of consumption and the lives of the colonial middle class.

Had Ajayi stayed in Nigeria, the attainment of the lifestyle that awaited him would have been incomparably harder. Not that he had been doing badly within the limits of his education; like other members of the C&S in London, it was relative success that fuelled his ambition, rather than failure. Some members describe depressing periods without work before they came to London, when they were competing with large numbers of similarly qualified young men in a saturated labor market. But for the majority of men it was their prior employment that enabled them to travel. A sample of 32 male elders revealed that 15 men had held clerical, secretarial, or minor executive posts, and eight were technicians or engineers. There were six government officials, two teachers and one male nurse. These men were the clerks, technicians, and minor officials with some secondary education who typified C&S male members in London. Often, like Ajayi, they were migrants into cities from villages or small towns who moved into the modern sector of government bureaucracy, services, and state and private enterprise established during the colonial regime. At this time, the urban sector accounted for only 5 percent of total employment, so they had already distinguished themselves from the great majority of Yoruba who were engaged in agriculture and trade.

Thirteen of the sample elders' fathers were in urban white-collar employment, as clerks, technicians, and minor officials, with one in the church and one a teacher. But 19 remained in the more traditional, rural sector of the economy. Men of this generation had limited opportunity for education; women had even less. A few of the fathers in this sample, like Ajayi's, had some secondary schooling, but most had primary only, and others had no more than basic literacy. Educational opportunities for their children had increased, but by 1947 there were still only 46 secondary schools in the whole of Nigeria.[4] The education that London students had already attained was therefore another measure of their relative success. The average male C&S member in London, like Ajayi, had already attained O level or the equivalent; Ajayi was not alone in looking for private tuition or study by correspondence course for O and A levels, City and Guilds qualifications, commercial certificates, or diplomas in an array of technical and administrative subjects. But these private endeavors in Nigeria still could not meet the students' ambitions. Western education was the only avenue into the ranks of the bureaucracy, but necessary credentials were hard to obtain in Nigeria. By 1968 there were still only five technical colleges, and the five universities mainly produced arts graduates, not technical and professional personnel.[5] Yet, at this point in Nigeria's economic development, access to the modern middle class was gained through employment in bureaucracies, either in large companies or, more likely, in the proliferating structures of the state. Anticipation of Independence in the 1950s and its realization in 1960 stimulated an explosion of professional opportunities as Africans moved into the privileged positions of departing Europeans.[6] By 1960, state boards and corporations covered all the major areas of administration and services; these expanded during the next decade, as did the civil service, while the division of the country into 12 states in 1967 further duplicated bureaucratic offices. Of a sample of the elite surveyed by Smythe and Smythe in 1960, 73 percent were employed by the state (1960: 78–79).

The alternative, employment in productive sectors of the economy, offered few such avenues into the middle elite, and into the lifestyle perceived as an expression of modernity.[7] Although there had been a proliferation of small-scale artisans, traders, and entrepreneurs during the 1950s and 1960s, commercial corporations were still largely in British hands; only very few Yoruba managed to accumulate the capital, credit, and skills to establish significant enterprises. Neither did the under-capitalized agricultural sector hold great allure for the upwardly mobile. The greatest farming fortunes, made mainly through cocoa, were already in the past, and even wealthy Yoruba farmers tended to be regarded as more traditional, distinct from the educated urban elite with whom students identified. Less than 2 percent of the Smythes's elite sample was primarily farmers (1960: 82).

So the structure both of the Nigerian state and of the economy privileged Western education as a means to power—and through recent colonial history, Britain was the obvious place to pursue it.[8] For students in London, framed certificates and letters after the name became a form of symbolic capital, with education seen not so much as a process, but as a prized possession, by which success or failure would be judged. "One thing you will find among all our people here, and that is ambition" said a member, "I will only go home when I have got the thing I came for." Teacher Ajayi expressed the same resolution: "We come to England for one thing only—to learn book and gain knowledge." To return to Nigeria without having obtained the "Golden Fleece" of qualifications was to admit failure and defeat, bringing shame both on unsuccessful students and their families.[9]

Worker-Students in London

The role model that Ajayi initially had in mind was an earlier generation of Nigerians in Britain, the "colonial" students who returned to professional preeminence, with lawyers the elite of the elite among them. These students had their equivalents in late 1960s and early 1970s Britain, studying at universities, the Bar, or medical school. Many came from educated or wealthy backgrounds, and family or state scholarship support afforded them fulltime study and the wherewithal to return home at the end of their course. Whilst in England, they were eligible for the benign attentions of the British Council and other welfare organizations for overseas students.

The members of the C&S in London were of a different order. Coming from nonelite backgrounds, with little financial backing, these worker-students had to pay their own way through college. Their dependence on their earnings to finance their studies is revealed by their years of entry, which was mainly prior to the mid-1960s (see table 2.1). Before this time, in response to the labor shortage in Britain, would-be students could enter the country as workers under a voucher system. But in 1965 a White Paper abolished vouchers for the unskilled, and limited skilled permits to certain professions. All

Table 2.1 Year of arrival of members of two male bands[a]

1940s	1950s	1960	1961	1962	1963	1964	1965	Post 1965	NA	Total
1	2	6	10	16	12	16	4	1	7	75

[a] These figures do not accurately reflect the actual numbers of members entering Britain in the early 1960s, since some had already returned to Nigeria by 1970. But the same trend is reflected in the British Council figures for all Nigerian students (British Council 1967: 27) and by census material on immigration from Nigeria (see Office of Population Censuses and Surveys: Country of Birth Tables 1974: 132–133).

would-be students were required to have prior confirmed college admission plus the finance for fulltime study. From mid-1962 to mid-1963, over a thousand Nigerian voucher-holders entered Britain; in 1967, just 20.[10] The door was closed to self-financing students. But for the worker-students who had beaten the ban, it was this contradiction between their identity as students and the necessity to work that produced the basic tension in their lives. Of the scores of C&S members I spoke to, none had enjoyed continuous fulltime study; all had, at some time, juggled employment and education in a variety of combinations. Although never abandoning their self-perception as students, the need to earn a living and support a family at times squeezed out study altogether, or reduced it to late-night stints with correspondence courses. All had therefore overstayed their anticipated absence of three to five years from Nigeria. Exactly half of the 1970 elders' sample was still in England in 1973. Of the 16 who had returned home, nine had been in England for ten years or more, two for nine years, three had taken eight years, and two, seven.

This distinction between types of students is necessarily crude; the interplay between the class into which students were born, that to which they aspire, and their educational experience in Britain is complex and defies tidy categories. But it is a contrast that resonates with church members themselves. As one male member said:

I never had anything to do with the British Council. We heard about it at home, but we never bothered with films and literature and all that. There is a class bias operating here—that thing . . . belongs to the educated people who are coming over to England to full-time courses and are attending college and university. . . . It's different for us; we are the struggling masses.

(Teacher Awofala)

Surviving on the edges of organized student life, the numbers of Yoruba worker-students were never counted. Census data shows 16,860 Nigerians in Greater London in 1966 and 18,540 in 1971.[11] Although most were Yoruba, these figures also include Ibo, Hausa, and other smaller Nigerian ethnic groups, as well as non-Nigerians born in the country. It also includes children, as well as diplomatic staff and those engaged in business. According to British Council estimates there were nearly 4,500 Nigerian students in London in 1969–70, the majority of whom are likely to have been Yoruba.[12] The courses pursued by men in the C&S reflected a combination of future opportunities in Nigeria with a realistic assessment of what they might achieve.

As shown in table 2.2, 8 percent of the male members' sample followed their predecessors' propensity for studying Law. Two did succeed, and by the mid-1970s were back in Nigeria. But the others either abandoned Law, or remained in England ensnared in the dilemma of being able neither to secure legal qualifications nor to return home without them. Two embarked on businesses whilst nominally still studying: one started a removal firm, specializing in transport to West Africa; the other became a representative for a Nigerian firm, traveling to and fro between London and Lagos.

But for Ajayi and other would-be lawyers, pragmatism tempered original ambition, as they adjusted to becoming senior technicians in their newly independent state. Well over a third of a sample of male C&S members in the early 1970s was, like Ajayi, studying some branch of Engineering. Accountancy was the next most popular subject, pursued by 17 percent of the sample above (although sometimes proving to be a euphemism for book-keeping or a HNC in Business Studies). Seven percent were engaged in Company Secretaryship with a further 7 percent in Quantity Surveying. Other courses taken by male church members outside this sample include Catering and Hotel Management, Printing, and Marketing. The qualifications sought included both relevant diplomas and certificates, together with membership in professional associations. The best of the professional

Table 2.2 Courses of study of 92 of the male members' sample, 1970 (in percentages)

Engineering	39
Accountancy	17
Law	8
Company Secretaryship	7
Quantity Surveying	7
Banking	3
Management Studies	3
Insurance	2
GCE	2
Plumbing	1
Advertising	1
Photography	1
Economics	1
None[a]	3
Not answered	3
Total	100

[a] The two Yoruba in this category were slightly older than in norm; one was a practising GP, the other a property owner. The third was a West Indian with a small business.

qualifications was on par with a B.Sc. degree, but only one member of the sample, the economist, was actually enrolled in a degree course; in this he was atypical of C&S members.

The quest for qualifications often led worker-students to attempt several courses at once, to switch from one course to another, or to add on evening classes or postal tuition in extra subjects. In practice this often overloaded the student and resulted in lack of focus. In many accounts of their experience, the search for a suitable course clearly revealed the parttime student's plight of being adrift in an uncertain situation with little opportunity for guidance. Those who came here under their own auspices without a place at college were more likely to discover that their previous qualifications were inadequate or inappropriate for the course they wished to take. When they did secure a place, parttime nonuniversity students suffered poorer supervision and teaching than fulltime undergraduates, contributing to a greater risk of failure.

Many were also not prepared with the necessary academic background, study skills, or proficiency in English appropriate for further education. Interrupted secondary education, with years of employment between school and arrival in Britain meant that Yoruba worker-students tended to be considerably older than their British counterparts or other overseas students. The majority were already in their thirties by 1970, and so had families to support. This contributed to financial problems: throughout their stay in Britain, the financial state of many worker-students remained precarious, often reaching private desperation. Personal savings for the fare to England may have been augmented by kin and community, but these contributions were not sustained. On the contrary, C&S members, such as Ajayi, were commonly expected to educate younger members of the family or send money home to ageing parents.

Yet the level of earnings that most members could command rarely left any slack for extra expenses, and employment was an area of difficulty regularly mentioned by C&S men. Their problems were commonly attributed to discrimination: "If an African wants to do television, they will teach him how to mend it not how to put on programmes." Another male member commented, "They will only give us the horrible jobs . . . Do you need a BA for sweeping?" Some members claimed to have noticed the NC (no colored) on the corner of job vacancy cards in the employment exchange or to have been refused interviews on the grounds of their Nigerian accent on the telephone.

Because of the disparity between the level of employment they hoped for and that which they achieved, worker-students often would not readily admit to the type of work they did, but low-paid jobs in the productive or service sectors were typical: "the office" might turn out to be a factory, or the packing basement of a department store. As an experienced sanitary inspector, Ajayi thought he had a good chance for an advertised post as an assistant public health officer with Westminster Council, but he was offered office cleaning. When he was a qualified engineer, he found himself on an assembly line. Another senior elder in the C&S had been a stationmaster in Nigeria, but in England he was a platform guard. The frustration of being offered work incommensurate with their expectations was not only due to the temporary nature of their employment; Yoruba also fell victim to the disadvantage of black workers in the shrinking labor market of the time.

Working Wives: Yoruba Women in London

For female members of the C&S, the situation was even more difficult. Women suffered discrimination not only in terms of their class position in British society and color, but also because of their gender. Black women's work in this period was concentrated in the light industrial and service sectors, characterized by lack of security with poor pay and conditions.[13] Female members sought work in factories, canteens, and clothing workshops. An alternative was exploitative outwork, sewing garments at home. Because of the chance to combine work with childcare, many female members spent endless isolated hours at their machines for a few pounds a week: in 1971, a dozen men's ties, for example, earned just 4/- (20p). Typing was also done at home, until a secretarial qualification offered the opportunity for office employment.

Taking her status largely from that of her husband, a woman usually worked to support the family and to help finance her husband's education, for her own education was rarely her initial reason for coming to Britain. Of the wives of the 32 men in the elders' sample, 27 initially came to join their husbands. Of the five who married here, one was African-Caribbean, one (who had already divorced her husband by 1970) was English, and a third had originally come over from Nigeria with another man. Only in two cases were qualifications their primary motive for independent travel; one, an older woman, trained here as a nurse; the other, the only one I knew to be financed by her wealthy educated family, came aged 19 to study Medicine, but moved on to Nursing, Secretarial Studies, and then Television Production, before settling for Catering, in which she qualified.

Nevertheless, once here, the majority of women wanted to use the opportunity to secure their own qualifications. Their ambitions had to take into account their previous educational level, which, because of

Table 2.3   Main courses taken by wives of
the 32 men in the elders' sample, 1970

Secretarial/Business	10
Dressmaking/Machining/Upholstery	8
Catering/Domestic Science	5
Nursing	3
Hairdressing/Wigmaking	3
Law	1
Librarianship	1
Interior Design	1
English classes	1
Not ascertained	4
Total	37

their gender, tended to be lower than that of their husbands. Universal primary education was only introduced into Western Nigeria in the mid-1950s; girls remained in a minority for the next decade, and formed an even smaller proportion of secondary school students. C&S female members in their thirties and forties might have had a year or two of secondary schooling at best, and still, by 1970, there were very few women in higher education either in England or Nigeria.

Other courses taken by female members included Advertising, Marketing, and Photography, as well as occupational training for semiskilled jobs such as computer punching or machine operation.

The total number of courses in table 2.3 does not add up to 32 since several women included more than one of the courses that they had taken while in Britain. The figure should be higher; from my experience, more than 10 percent of women had at some time followed Secretarial or Business Studies, which would prepare them not only for bureaucratic employment, but also for a business or trade. The second most popular category, sewing (which is also certainly underrepresented in table 2.3), could be combined with commerce: one female elder outlined her plans to me for a future business in dressmaking and imported cloth. Nursing also guaranteed employment on return to Nigeria; several women in the C&S were training for qualifications at various professional levels.

Marriage and Mothering

Many female church members succeeded in returning home with qualifications. But whilst in London they were caught in a triple bind: not only did they have to manage both study and employment, but also the disparities between their own agenda and their role as wife.

The hope harbored by many women was that their greater earning capacity would give them more leverage inside their marriage, and greater security if it failed. But in spite of the skills they were acquiring, women still felt at a relative disadvantage to men in a community where education was so highly valued: "Women must muddle up because they are not much educated," said one male member. Moreover, *pace* Engels (1958) and the orthodox socialist line on women's liberation, women's economic participation does not necessarily entail equality. In Nigeria, an independent income from trading, in which women commonly engaged, had not led their mothers to challenge men's authority; women in London with whom I discussed gender relationships accepted that men were, as they put it, "the master," the household heads, to whom wives owed obedience. Women could grumble about their respective husbands and joke irreverently about men, yet they still attempted to negotiate the demands of their different roles without challenging male dominance.[14]

The demands of life in London also upset the shared cultural expectations of a domestic relationship while offering no alternative model to emulate. This conflict was not necessarily apparent: I often saw men preparing the soup, the red-pepper stew that would be eaten with rice, yam flour, or other starch accompaniment throughout the week. Husbands would shop in the Brixton and Shepherd's Bush markets and haggle over live chickens with the poultry dealers in Whitechapel; they would bathe and look after the children—chores that a man would rarely undertake in Nigeria. In the absence of kin, with both partners working and studying, many women pressed for this sharing of domestic work: "You only have yourselves [i.e., each other]. If you don't help yourselves, no one else will," said both wives and husbands. But in practice male resentment could fester.

Naturally, the individual relationship between husband and wife varied greatly, but this clash of expectations in a precarious environment often exacerbated problems. What might be tolerable in the accepted norms of marital behavior in Nigeria could become unbearable when in London. In theory, for example, women accepted their husband's freedom to come and go as he pleased, and to socialize without the family. But, denied the same liberty, and deprived of the company of extended family and a close circle of female neighbors, wives often had to endure long evenings on their own. Although a few older male members had left a first or less educated wife at home, none maintained two wives in London.[15] But the knowledge that their men were "going all about" without them gave rise to women's apprehensions of their husband's infidelity. When discussing the difficulties of married life in London, women often mentioned jealousy and loneliness as crucial factors.

These pressures combined to place many marriages under strain, and some couples in the church during my fieldwork did separate. One precipitating cause for this, as with the Ajayis, was the failure to have children. With Western notions of companionate marriages in mind, I once asked a member whether a couple he knew well was happy together. He answered: "She has two issue [i.e., children] now, so why not?" Although it might be expected that urban elite couples would desire smaller families, research undertaken in post-Independence Nigeria showed that they still hoped for several children, even though they enjoyed lower rates of infant mortality and were not bound by the constraints of producing family labor, as would be their rural relatives.[16] Failure to conceive was a calamity for both the wife and her husband, and a valid reason for divorce.

Yet the arrival of children brought further problems. Although the onus for fertility and basic childcare lay with the woman, the conflict between their productive and reproductive roles affected both partners. In the absence of the extended family, worker-students in London had to combine career ambitions with parenthood. One solution was to send the children back home to live with kin, but parents worried over inadequate medical attention in Nigeria, a fear that was tragically realized in the death of one child being cared for by grandparents. Members who kept their children with them in England competed for scarce places in a day nursery or looked for a private childminder. But minders were often unregistered and unsupervised, and standards of care erratic.

An equally popular strategy was to send the children away, not to Nigeria, but to be fostered with white, working-class families, who often lived outside London.[17] This delegation of responsibility was not an unfamiliar pattern. In Nigeria, Yoruba children often live with kin or some other known and respected family, both for their own education and training and to assist with domestic or productive work. Many members had themselves been brought up apart from their own parents, absorbed into a wide network of relatives and neighbors. In Britain, the social context was completely different. Not all children of church members in Britain had negative experiences of fostering, but some children had lived with several different families in the space of a few years. This lack of continuity compounded a potential conflict round their own identity in terms of language, culture, class, color, and emotional allegiance. Confusion could be expressed through behavioral disturbance, and difficulties with children and childcare arrangements caused much anxiety among Yoruba families in London.

Finding Accommodation

A contributing factor to childcare problems was the overcrowded conditions in which many families lived. The localities in which worker-students looked for accommodation were in areas of mobile, low-income populations, including black immigrants, living on the margins of the city. The terraces of dilapidated Victorian houses typical of Stoke Newington, Hackney, Finsbury Park, Brixton, and Shepherd's Bush had not, by the early 1970s, undergone middle-class "gentrification," and it was in these subdivided, multi-occupied dwellings that many members were housed. In the second half of the 1960s, only 3 percent of the British population rented furnished accommodation.[18] By contrast, 70 percent of C&S elders in 1970 were renting, and among younger, more recently arrived members, the proportion was even higher. Although Nigerians were assigned to the same housing sector as many black working-class families, research carried out in this period indicates that in their comparative level of multi-occupation, overcrowding, and the sharing or lack of facilities, West Africans landed at the bottom of the heap.[19] Their experience, as they described it, was often subject to the attitudes of white landlords to black room-seekers and the inequitable treatment of black tenants regarding rents and conditions.[20] The minority of black landlords was not necessarily more generous. It was "landlord trouble"—harassment, insecurity of tenure, and fear of eviction—that precipitated Ajayi's frequent change of address and often brought anxious members to the C&S for prayer.

They also were at a disadvantage in the second option for housing, the public sector. In 1966 only 4 percent of the black population in London were council tenants, as compared with 22 percent of natives. The figure for West Africans was 2 percent.[21] Of the 32 members of the C&S elders' sample, only one had a council flat in 1970, though three more had moved into local authority property by 1973.

A third option was private ownership. House prices, especially for run-down properties in marginalized areas, were not what they are today. Although not representative of members as a whole, 9 out of the 32 men in the elders' sample had purchased their own properties. A downside of ownership was the impression of wealth and success conveyed to family at home, with the expectation of largesse this entailed.

Members often complained to me about the misconceptions their families in Nigeria entertained as to the living conditions in England. "Furnished flat" could be a euphemism for a few pieces of chipboard furniture in a partitioned room. Facilities were often communal, with shared bathroom and lavatory, and one cooker on the landing to be used by several households. Overcrowding was but one of the hazards; the material state of much of this housing also exacerbated social tensions, reduced opportunity for effective study, and endangered physical and mental well-being. When I first visited one senior elder, for example, he was renting a tiny dark room backing onto the railway line in Brixton's Railton Road, since demolished. The door could hardly open for the bed, which was squashed beside a table and a chair; the dirty wallpaper was bubbling with damp. The one bath, unusable for years, was full of rubbish; the lavatory was cracked, with the walls sprouting an impressive array of fungi. The shared kitchen was infested with mice, and in total disrepair. When a Yoruba neighbor upstairs let her sink overflow, his ceiling collapsed, and he was forced to move on.

For some C&S families, living, sleeping, and sometimes cooking too, had to be carried on in one room. Behind a curtain suspended from a wire, an old-fashioned double bed jostled with the chest-of-drawers and wardrobe, on which would be piled suitcases and boxes storing a woman's collection of *iro* and *buba* (wrappers and blouses) and the items steadily accumulated in preparation for return. In the front of the room, there might be a three-piece suite, radiogram, and coffee table with a decorated plastic top. For C&S members, the table or television often doubled as an altar, with candlestick, crucifix, Bible, and a bottle of Holy Water. Over this could hang a picture of Christ with the Bleeding Heart, together with a few calendars, a wedding photograph, or pictures of family and friends at home. In winter the room would be filled with the damp warmth of a paraffin heater; in summer, the curtains closed against the sun, as they would be in Nigeria.

For "home" continued to provide a critical cultural reference for Yoruba worker-students. Their rooms rarely acquired the sense of permanence reflected in some West Indian households, with their profusion of ornaments, doilies, and cabinets of china and glass. The attitude of Nigerian students toward their housing was pragmatic rather than aesthetic: insecure and inadequate accommodation in London was yet another "obstacle" to their return.

Class, Race, and Identity

For Yoruba students, their housing situation perfectly illustrates aspects of their predicament. The geographical areas and sectors of the housing market in which they lived reinforced their position as part of the proletariat, as did their position in the labor market. They saw themselves as part of an educated elite, but their self-construed identity was "locked up," as Elder Oguntulu put it, by their daily experience. The notion of identity is a complex package. At its most basic it refers to "who we are, 'where we're coming from.' As such it is the background against which our tastes and desires and opinions and aspirations make sense" (Taylor 1994: 33–34). This process of self-definition entails the articulation of two dimensions. The first is temporal, a narrative connecting a present sense of selfhood with both past states and future aspirations.[22] The second is spatial—the interrelationship between points of similarity and of difference between self and contemporary others. Identity also refers to the relationship between how people see themselves and how they are seen.[23] Interpretations that highlight self-construal understand identity as a fluid process rather than static status. In the case of transnational migration, Rouse (1995) emphasizes, part of this movement is between the multiple identities that migrants maintain, rather than sustaining a single, fixed self-definition.

All these implications of identity illuminate the dilemma of C&S members, for each aspect was conflictual. Through their odyssey in England, Ajayi and Oguntulu not only equipped themselves for well-paid employment in Nigeria, but also entered the Yoruba middle class. Although this class transition was only fully realized on return to Nigeria, the equation that Yoruba students made between higher education and superior status encouraged them, while still in London, to rank themselves among the elite, eliding future success with the process of attaining it.

For Yoruba themselves, the category of "student" signaled seniority and status. But in contemporary Britain it stood for irresponsible youth. Gaining little standing in British society through attending college, C&S members received even less as workers. They were aware that the average Briton would not only assume that they were part of the labor force rather than students, but also that they were working class. Their self-conception as professionals was not reflected back to them by the wider society; on the contrary, it was denied. Engaged in manual work, living in run-down areas of the inner city, they were, to all appearances, part of the proletariat.

This ascribed identity was not just a matter of class, but also of race. Yoruba worker-students not only had to negotiate white culture, but their status as black "immigrants." Because of their color, they were commonly assumed to be part of a black immigrant under class, lumped together in popular perception with what were then called

"West Indians."[24] Many church members expressed bitter resentment at this confusion, which compromised their own middle-class identity, and, as they saw it, their superior racial origins. Some went so far as to voice support for the Conservative MP Enoch Powell in his call for repatriation, maintaining that these policies did not refer to themselves, but to *ireke*. This is a Yoruba term for sugarcane, an epithet that implies slave and was used as a derogatory term for African-Caribbeans. In return, I heard West Indians refer to West Africans as "savages." In this period neither popular culture nor politics diluted mutual distrust between the two communities. Whilst reggae was capturing white and black audiences in London, Nigerian juju and highlife music had not yet hit the clubs, and records were available only in one small shop in Warren Street.[25] It is ironic that while British Rastafarians were mythologizing Africa as the Promised Land, with elements of African culture assuming iconic significance for their political struggle, the representatives of this Utopia who were living in Britain held no significance for the movement.

Nor did radical African-Caribbean youth court Yoruba students for their cause. The late 1960s and early 1970s were years of burgeoning black militancy; inspired by the Black Power movement in the United States, African-Caribbean youth confronted the authority of the state both through political organization, and on the streets. But Yoruba worker-students tried to avoid being tarred with the brush of black activism. Despite their experience of racism, not only did worker-students avoid participation, they positively rejected black politics and a political identity based on color. Their own political views were those of an aspirant national bourgeoisie, not of a black immigrant proletariat. But this lack of political involvement was not only related to a desire to distance themselves from the African-Caribbean community. Seeing themselves as transient migrants rather than settlers, they situated themselves in a different relationship to Britain from both first- and second-generation black radicals: "West Indies people are so aggressive because they think they have 50–50 rights in England" said Teacher Korode. This is the voice of a temporary resident; it also speaks for a community whose national identity and locus of interest lies elsewhere: "Our only job here is passing examination; we ignore all the rest."

The desire of Yoruba worker-students to keep a low profile was also pragmatic. To become involved could have jeopardized their education; to draw attention to themselves invited investigation by immigration authorities, and, if irregularities emerged, the threat of deportation. Church members preferred public invisibility to a political presence and expressed little confidence in the efficacy of protest. The sense of strength that comes from common action, the solidarity experienced through organization, came from membership of the C&S rather than political support of black comrades; when racially abused, members turned to prayer rather than collective action. In 1971, an elder was attacked on the underground; in defending himself, he slightly injured one of his four assailants, and was subsequently charged with assault, found guilty, and fined. I asked him if he would appeal: "No, I would never cause more trouble like that; I want to stay out of trouble, not get into it. Anyway, I don't think that the law can solve any problem. . . . There is no one to speak up for me in this country."

Being popularly confused with African-Caribbeans, Yoruba came in for the discrimination encountered by all black communities. Racism directed at any one ethnic category has two aspects. First, if there are negative images of other groups who are perceived to be phenotypically similar, or who share a social or economic position in the social structure, the category in question will be also be a target for this generalized hostility. But second, racism has its specificities, constructed around the particular history of the group and its relationship to the dominant society. In members' experience, if they pointed out the distinction between themselves and West Indians, this did not produce a positive reaction, but only summoned up a different set of stereotypes. Their work mates would ask whether they knew how to read a newspaper, whether they really lived in trees, or whether people in Africa have tails?[26] C&S members commonly responded with irony: "No, but I like looking at the pictures"; "The Queen was in Nigeria in 1956; why don't you give her a ring and find out?" They claimed to brush such rudeness off as ignorance, and affected a defensive indifference: "I don't worry about petty abuse, only when it obstructs what we come for." But the insults still astonished. How could the British know nothing about a country that was once part of their colonial heritage and is now a partner in the Commonwealth? And why the hostility and ridicule?

What was for Nigeria the dawn of Independence was for Britain the end of empire. Postwar confidence was over; the economic security of the 1950s had evaporated; by the end of the 1960s the economy was sliding into the recession that was to characterize the following decade. Labor shortages were replaced by structural unemployment; clashes between working-class youth and the police during the 1960s and early 1970s drew attention to inner-city decay, and fuelled fears about declining law and order. In this situation, the concept of race served both to account for and control this perceived decline.[27]

The actual size of black and Asian communities in Britain was very small, still only some 2.5 percent of the total population by 1971,[28] but its symbolic significance was enormous. As a second generation swelled the numbers of black British, and as ethnic communities assumed distinctive identities, the black minorities were held responsible for national ills. In the popular media their numbers and alien cultures were not only held accountable for consuming scarce resources, but actually for creating the poor conditions of the areas in which they lived. The Conservative MP Enoch Powell's notorious prediction of "rivers of blood" was made in 1968, the year before I started fieldwork, and served to legitimate popular racism in Britain.[29]

Circumstance and Choice

To portray Yoruba Aladura in London solely as casualties of circumstance, as passive, nonmilitant recipients of racism and its attendant problems is, however, to distort their experience. They were agents as much as victims. Their dilemmas did not stem solely from British attitudes and institutionalized inequalities; they were affected also by personal choices: short-term advantage in employment, housing, or childcare, for example, might be deliberately sacrificed for future gains.

Agency also takes culturally appropriate forms. The church was not overtly counterhegemonic; the members were not seeking an indirect route to political influence. But they were looking for empowerment in their own terms, a personal strengthening through spiritual efficacy that was thought to have direct bearing on their lives. It was anxiety over their problems—whether of childcare, education, employment, housing, immigration, racism, or family life—that brought members into the C&S; it was the difficulties they encountered that took them to C&S prophets for prayer. In the Aladura cosmology, this constitutes pragmatic initiative, not passivity.

It was also their church that helped them mediate the dissonance between their position in British society and their self-perception. A consideration of C&S rhetoric corrects a Eurocentric interpretation of worker-student experience as one of unmitigated disadvantage. C&S discourse portrayed the church not only as a refuge from present indignities, but also as an association of the successful.

Cherubim and Seraphim is different from other churches, because in Methodist hardly can you see a professional man or woman. But in Cherubim and Seraphim at home, in this organisation they don't use money for anything because the Lord will call the lawyers, we've got the doctors, we've got the accountants, we've got the barristers—all various things—the Lord will call us to meet together . . . to work in his own vineyard

(Sermon 1970)

Constantly reminding them of their achievements and ambitions, the C&S helped members see themselves as upwardly mobile rather than as coming down in the world in London. As well as addressing their present problems, it held out a picture of the rewards, and their glittering future as "Stars of the World."

Chapter 3

The Cherubim and Seraphim Church, United Kingdom

Going Home: The Adisas' Thanksgiving Testimony

Brother and Sister Adisa face the congregation, looking down the long blue-and-white painted hall in East London, over the pews of some hundred white-gowned men on their right, and the veiled heads of an equal number of sisters on their left. At the back, a group of their friends who are not members of the church sit among the visitors, the brilliant blouses and wrappers of the women crowned by stiff damask headties, which point up over the sea of prayer gowns. On a dais behind the Adisas, the chairs of ten male elders are ranged on either side of the modern altar. On the wall above, the painted inscription "Holy, Holy, Holy, Lord God Almighty" curves over a large wooden cross, roughly wired with red strip lighting, and flanked by two fat cherubs representing the Cherubim and Seraphim.

Amid the noise of children moving around the pews and the shuffling of babies, the congregation listens attentively to the couple who are standing where they all hope to stand, offering their final thanksgiving:

> Brothers and Sisters, we thank the Lord, and want to let you know how far the Lord has helped us. For those that are just coming here now, who were not with us say around 1967, I must just say a bit about myself.
>
> It was my blackest year, and the year when I nearly lost my life, having lost both my parents. The Devil was waging war, and there were so many obstacles in my path . . . I lost all hope; I thought it was the end of my life. But then I got to know more about this church. Of particular prominence was Prophet Fakeye who used to come to us in the night

to encourage me, and tell me of his experiences too, before he came to this country. We have had our ups and downs, but we have been able to stand the stormy weather through your prayers. I must mention here . . . those people who have been acting as sort of spiritual guideposts. Certain committees or some meetings which we go to, like Choristers—any time my wife went there, she would come home with a lot of advices [i.e., through revelation] to guide us through. And to have been guided all the while by people like that . . ., [they] have shown us what practical Christianity is.

You see, it is nice to take your failure and success with equanimity. I mean by saying this that failure and success often come . . . and when trouble strikes we tend to stand and think, "O, this is the end of the world." And then you want to take off your white robes away and keep them [off] for ever. That is when the Devil succeeds. But if you refuse, and still keep steady, then you will have achieved what you want, and this is what we have enjoyed by the Grace of God. . . .

We will now be going home for better, because God has provided a job for us, and not only that, but a lot of other amenities have been provided to make us comfortable, and we thank God for this. This is what you have done through your prayers, and I thank everybody here who have been instrumental to this successful end to our journey to Great Britain . . . .

May God bless you, and keep you and be with you always, and make this church a beacon of hope for unbelievers. Thank you very much.

(Service 16 November 1969)

The elders to whom Adisa pays tribute were seasoned members of the C&S from Nigeria, but the majority of the men and women listening to this testimony had, like the Adisas themselves, joined the church in England. Of the 40 men on the register of one male Band, only 13 had been members "right from home." But both through the constant reference to the founder in prayer and sermon and the high profile of the C&S in Nigeria, they perceived their religious organization as Nigerian rather than as a British church.

The Nigerian Origins of C&S

No Yoruba could fail to know of the C&S in Nigeria,[1] through casual visits, through friends and family who attended services, and from the familiar sight of white-robed followers parading through the streets, praying on hilltops, and prophesying on Lagos beaches. Estimates of members in the Western Region during the 1960s vary from 50,000 to 80,000 but the number of those consulting the prophets in private was far greater. At that time, the C&S was the largest of the Aladura churches, so called because of their faith in the absolute efficacy of prayer. By 1970, the C&S had gained a high profile throughout Nigeria, and was well established in Ghana, Sierra Leone, and Benin.[2] The church had its origins in a Lagos prayer group established by Moses Orimolade Tunolase, who was both disabled and illiterate, and a young Anglican woman from a well-to-do family, Abiodun Akinsowon. Initially, their followers did not plan to break with the mission churches they attended. But the spontaneous fervor of their prayer and possession by the Holy Spirit proved too much for orthodox decorum, and in 1925 they were forced out of mainstream congregations into independency. A C&S member in London described to me how in the late 1920s his uncle returned from trading in Lagos as a Christian convert, and enthusiastically joined the local CMS. "He used to be standing up there saying visions to the people. He didn't mean to split off, he meant to bring something in. But they shouted him out." The founders of the first Aladura church—that which was to become the Christ Apostolic—had already suffered the same treatment; another Aladura group, the Church of the Lord, was to follow.

The subsequent history of the C&S in Nigeria is a minefield of splits and secessions.[3] The stage was set by Abiodun herself, who, frustrated in her aim to head the church, abandoned Orimolade's Eternal Sacred Order of the C&S (ESO C&S) in 1929 to lead her own branch, later named C&S Society. The next defection in the following year was of The Praying Band, itself to divide in 1932. Several of the disputes in C&S history center on the claim to the position of Baba Aladura, "Praying Father," or head of the church. By the end of the 1970s there was "more than a score of Baba Aladuras, two-third of them based in Lagos" (Omoyajowo 1984: iii); by 1968 there were already 14 legally registered sections of the C&S in the capital alone, most of which had a ramification of branches throughout the Western Region.[4] Then there were the numerous smaller groups, which had generated their own offshoots, besides the hundreds of unaffiliated local congregations and independent prophets all identifying themselves as C&S.

Although this process of fission helped to spread the church, internal squabbling diminished its reputation. In spite of the image that members in London portrayed of the C&S, it did not include many of the top Nigerian elite, who were more likely to seek private consultations with prophets than openly identify with "white garment" groups. But the wide popular support was not, however, confined to the rural and urban poor who gravitated to Aladura in response to the social and economic uncertainties of colonialism. As Peel emphasizes

(1968: 83–91), it was those taking advantage of new opportunities—clerks and artisans, as well as converted farmers and traders—who particularly responded to this version of Christianity and formed the backbone of the C&S as it developed.

A characteristic of this "marked and distinctive type" (216) was its geographical mobility, as well as its social ascent. Many Aladura were migrants from rural areas seeking a new life in the towns. Some had gone further afield, to northern Nigeria. The predominance of Islam amongst the Hausa and other northern ethnic groups meant that Western education had lagged behind the rest of Nigeria. The development of Kaduna, Jos, Zaria, and other towns on the railway line during the 1920s therefore depended on clerks and government employees imported from the south.[5] As their numbers grew, those who had been members of the Cherubim and Seraphim at home started meeting together to pray and formed the Northern Movement of the C&S.[6]

From the start the northern church attempted to avoid entanglement with southern factionalism and to establish its own administrative structures, in which central control was combined with a measure of democracy. During the 1960s the establishment of a university and technical and commercial training centers in the north increased the proportion of skilled and professional men in the church. Their experience of staffing state and commercial bureaucracies provided them with models of administration, plus an incentive to save economically valuable time with efficient organization.

Among these young men were future leaders of the U.K. C&S. All of the men instrumental in establishing the London church had worked and/or studied in the north: the whole of the ten-man Executive set up in January 1966 were "northerners." What they brought with them was the experience of organizing a church among the Yoruba diaspora in northern Nigeria, to stand them in good stead when developing the C&S among the Yoruba community in London.

What we find, therefore, on comparing the London congregation in the second half of the twentieth century with its Nigerian beginnings, is continuity concealed behind apparent contrast. First, both sets of new recruits tended to be young, becoming Aladura before the age of 40. Second, both were immigrants, whether from the countryside into urban areas, or as part of a diaspora. For members in Britain, although their self-chosen exile was, they hoped, only temporary, they shared the same dislocation from their roots, the same separation from a familiar community and kin. Third, both generations shared the experience of class transition. To use terms loosely, many early members were drawn from the emergent petite bourgeoisie, the first generation in their families to receive Western education. The later Yoruba diaspora in Britain was a would-be elite, aspiring to bourgeois status through the further and higher education their parents lacked. Both were the middle classes of their day.

The Foundation of the C&S in the United Kingdom

The beginnings of the C&S in London followed the familiar pattern: the prayer group that evolved into a church. On, June 4, 1965, inspired by previous prayers between friends, a meeting was organized as both a spiritual and a social event. Some 25 people turned up, 22 of them members "from home."

The service program followed that afternoon was that of the Northern Movement.

> Senior Apostle Rose Osazuwa played his accordion at this prayer meeting and there was thunderous clapping of hands and rendering of songs and praises to God. The spirit descended mightily and Elders P. Korode and Ogundele went into trance. The Lord then spoke to the congregation through Elder Ogundele, interpreted by Prophet Fakeye,[7] and gave assurance that on that day was established the "Cherubim and Seraphim" Church in the United Kingdom.
>
> (Chirwa 1990: 10)

Developments followed fast; an Executive committee was formed, which decided to advertise in the weekly magazine *West Africa*. By the beginning of August the C&S Movement, United Kingdom, had rented King's Weigh House Church, Binney Street, in central London. A year later, some 500 people, 200 of whom were regular members, attended the anniversary.[8] The following year the C&S moved to Oaklands Congregational Church, Uxbridge Road, in Shepherd's Bush, and by 1970 the bimonthly Sunday services had congregations of around 350 adults, with between 100 and 200 children. The church was at its peak.

After the 1971 purchase of a building in Earlham Grove, Forest Gate, E17, attendance at services averaged 200 adults, and had dropped to around 150 in 1973. By this time, many of the members had returned home; only Abidoye from the original 1965 Executive was still in London, and, although new members were constantly joining, the body of potential recruits was dwindling. There already existed by this date the nuclei of competing C&S congregations (see chapter 9), and, as Sunday services took place weekly, with spiritual

"Watchnights" on alternate Saturday nights, only the most assiduous attended every service. Nevertheless, the acquisition of its own hall was critical to the development of the C&S in Britain, to its recognition as an established church for registration as a charitable organization and eventual acceptance by the British Council of Churches in 1978—the first black-led church to obtain full membership.

The C&S as a Yoruba Association

The purchase of its own property enabled the church to expand its activities. The Sunday service was only one site for the social contact generated by ritual: Adisa's testimony touches on the multiplicity of sources of potential support, including the Bands, the four male and four female subgroups of the congregation, which members were encouraged to join for monthly prayer meetings.[9] Adisa also pays tribute to the core of C&S practice, the constant comings and goings for the purpose of prayer. As C&S prophets and visioners became established, the Bands appointed officers to visit at home; a troubled member could call on others for assistance at any time of day or night.

Then there were the social occasions. The Adisas' send-off service was followed by yet another party in their honor. For the annual anniversary of the church and each Band, halls were hired, DJs or sound systems installed, huge pans of rice and pepper stew prepared, and crates of soft drinks hauled in. Although condemning smoking and excessive alcohol, Aladura did not espouse rigid codes of conduct; in recreation too, they aimed at effective participation in society, rather than rejection of it.

The style of these celebrations, as with members' private parties to mark births, marriages, and other life events, was Yoruba: the seating of guests, the speeches and welcomes, the music, dancing, and dress. It was the reinforcement of Yoruba conventions that was underlined at these occasions, not an incorporation of British culture. It was also a Yoruba social network that was strengthened, one that could be called upon not only for celebrations, visiting, and as a clientele for commercial ventures, but in the serious business of finding housing, childcare, and employment. Church records include references for jobs and colleges, evidence for immigration authorities, together with letters of recommendation for members returning to Nigeria. The C&S provided not only a potentially supportive social web in London, but a means of maintaining a Yoruba identity, Nigerian contacts, and an orientation toward home throughout the protracted stay in Britain.

It also countered the loneliness of London life. In all previous descriptions of Nigerian students in London, whether fictionalized or ethnographic, it is this that emerges as their common scourge.[10] For people accustomed to the communality of compound, village, or town quarter, British privacies seemed shocking, and isolation became a torment. The social void intensified anxiety, and the psychological pressure of studying could assume somatic force:

> Sometimes when I came home at night and I begin to think, the loneliness starts to get into my brain. I feel my head swell up and I thought it would burst. There was no one around to talk to and no one to turn to. It was very terrible.
>
> (Teacher Oluwole)

Ajayi was not the only one to sink in the "quicksands of agony." The fear of what was known as "brain derailment" was widespread, and well founded: a study undertaken in 1968 in the London Borough of Camberwell found that the rate of mental illness among Nigerian students was four times as high as among the native population,[11] a conclusion corroborated by other research on the mental health of Yoruba students in England. For many members, the C&S did more than stave off disaster. Pastor Tomori was one:

> When I arrived in 1962 there is no church yet. Many students were arriving, and some get so lonely, frustrated, dejected and despondent that they gas themselves. Others are failing their exams repeatedly, and once your heart is broken you fall sick and have no interest in living. But the church has made a great difference. It is a social place where you see friends—you dance and clap and are happy together, and you forget all our worries . . . You are meeting people and are having things in common with them. It cheers you up and people get life under their age [i.e., feel younger than their years].

The common interest in the church could override regional origins. Yoruba ethnic subgroups assumed significance in friendship and patterns of allegiance in church politics, but the C&S acted as a pan-Yoruba association in the diaspora context of London, as it did for urban migrants in Nigeria itself.[12] This was strengthened by the ideology of kinship promoted by the church, which appropriated aspects of the absent extended family. Brother, sister—the idiom of kinship among members carried connotations of the rights and obligations associated with the Yoruba lineage and household. For women, female church associates stood in for relatives and co-wives with whom they could socialize with their husband's approval. Despite the acknowledged danger of "C&S Casanovas," men would allow their wives to attend C&S occasions unaccompanied, a freedom that women greatly valued.

In the style of other African independent churches, the C&S also intervened in matters customarily handled by parents or senior kin. C&S representatives, for example, were often called upon to make peace between feuding members, and settle disputes, both domestic and otherwise, according to Yoruba norms of mediation. Members might use this procedure to avoid exposure to British courts: a husband negotiated access to his children when his ex-wife gained custody; a landlady sought help with uncooperative tenants.

But the C&S did not command the same clout as the customary authority of locality or lineage; it was not a straight substitution for kin-group control, already loosening its grip on those living away from their closest family. Apart from social pressure, the elders had few sanctions at their command and needed the voluntary cooperation of members. Unless the matter concerned internal C&S affairs, such as the censure of inappropriate behavior or deviations from ritual practice, the elders acted only on request and could do little if the parties rejected their intervention. The C&S thus negotiated a path fitting for a voluntary association in a changing social context, between upholding ideal Yoruba morality and leaving members to their own pursuits. C&S practice allowed room for the personal choice appropriate to a would-be bourgeoisie, whilst offering support from the perspective of indigenous conventions.

Spiritual and Administrative Authority

The space for the exercise of individual agency was also reflected in the organization of the church. As with the bureaucratic forms that were becoming entrenched in secular Nigerian society, the membership of C&S was organized into a hierarchical structure through which everyone could ascend.

For men, the spiritual hierarchy consisted, as it does today, of ranked grades. The base line is that of "brother." Any man in the congregation is addressed in this way, whether he be a regular attender, an invitee, or what is known as a "floating member": a person who comes and goes, visiting the church with a particular problem, melting away again once circumstances improve. Regular participation for a year or two qualifies for ordination into the first grade of "aladura," or "prayerist." The next male grade is that of "elder," then comes the rank of "teacher," a role that implies a capacity to pass on the principles of C&S rather than a formal duty to do so. In order of precedence, the next grade is "pastor," with a single incumbent at any one time: "He must be the right hand man of the Spiritual Leader of the church to inform him of what is going on in the church" (Abidoye n.d. [1971]: 19). Then comes the rank of "evangelist."

The following grade is "prophet." As for the female counterpart, the "prophetess," the "job is mainly spiritual. To pray for people and say [i.e., reveal] vision . . ." (18). Prophets are among the most influential men in every aspect of the C&S, and are few. There were two prophets in 1968; two more were ordained in April 1970, but none at the next ordination of October 1974, even though all but one prophet had returned to Nigeria. The grade may also be combined with the final rungs in the ladder: "apostle," "senior apostle," and "special apostle," ranks that involve both spiritual and administrative authority. For women there are but five "spiritual" ranks, there being no female equivalent of teacher, pastor, or evangelist. The roles of "aladura sister" and "elder sister" mirror those of their male counterparts, as does prophetess, of which there was only one. But "lady leaders," although exercising recognized seniority in the church, were expected to focus their efforts on the female members, especially through Bands, rather than the congregation as a whole.

C&S pamphlets from Nigeria such as the ESO C&S *The Order* (n.d.) specify the duties and qualifications of each grade. But in practice these are conceived in general terms, allowing scope for individual enterprise—or sloth. Here the bureaucratic model falters; role definition is weaker than the principle of rank, for the primary criterion for promotion (apart from the posts of prophet and prophetess) is length of membership and evident commitment to the church. C&S doctrine is more about doing than believing, acquiring the relevant ritual knowledge and deepening understanding of C&S mores comes through repeated performance rather than formal instruction. Thus ranking is determined not by age but by "spiritual maturity," that is, involvement in C&S praxis. This experience, it was said, will develop the capacity for effective prayer, the raison d'être of the C&S.

By 1970 the number of ordained members had risen to 133, as compared with 62 two years before. This expanding spiritual democracy was a powerful attraction of Aladura for its members. For the majority, the possibility of attaining seniority in their church was a novel experience. Only about a fifth of male members had been born into Aladura. Most—over 70 percent—had belonged to mainstream denominations (mainly Anglicans, with lesser numbers previously attending

Methodist, African, Baptist, and Roman Catholic churches, in that order), and nearly 10 percent had been Muslims.[13] What they discovered in the C&S, as with other Aladura, was a church where authority, both spiritual and organizational, could be achieved by members rather than remaining the prerogative of minister or priest. Ritually, this possibility was demonstrated by congregational participation in worship; doctrinally, it was justified by a reformulation of the individual's relationship to God: "every member a priest." Although this formulation owes much to the Protestant tradition, the orthodox Christian priest or minister was privileged with respect to communication with the divine as purveyor of sacramental grace. He was also, in colonial Nigeria, the one who held knowledge of God's word through his understanding of the Bible. But Aladura challenged this clerical hegemony by asserting the capacity of every lay individual to communicate with God directly, bypassing these intermediaries. Every member has access to the power of the Holy Spirit, a circumstance that Legesse has labeled "democharisma," where "the bearer of charisma is not the prophet" but the "ordinary members of a charismatic community" (1994: 337). Through personal effort, and obedience to Christian morality, there was now, they proclaimed, immediate access to the source of power, and thus a new spiritual democracy.

It was women who benefited most from early ordinations: in the first year of the church there were 6 male and 4 female Aladuras; by 1970 the figures were 31 and 40 respectively. The number of elder sisters rose from 2 to 20 as compared to 7 to 17 for men. The number of women in these two grades thus increased tenfold over the first six years. Unlike orthodox practice of the time, women had a significant voice in revelation and prayer; they managed their own ritual activity in the female Bands, and could also look forward to ordination. There were some formidable female figures in the church.

But in general, the fusion of indigenous bias with patriarchal Biblical sanction resulted in the retention of male dominance in the spiritual sphere. Despite the conclusions in earlier literature on women's participation in African independency, neither C&S practice, nor that of other Nigerian indigenous churches, revolutionize cultural assumptions concerning gender.[14] In his assessment of "The Position of Women" in the Nigerian C&S, Omoyajowo (1982: 200) points to the significance of women such as Abiodun in C&S history, but suggests that Orimolade's cofounder was not recognized as the leader of the church "simply because she is a woman" (201), an accusation Abiodun herself endorsed in an interview (September 6, 1971). Although both Brother and Sister Adisa came before the congregation, the sister remained silent while her husband spoke on her behalf. In the hall where they stood, the lady leaders, prophetesses, and elder sisters sat to the left of the altar platform or in the foremost pews of the women's side of the hall; none was on the altar platform. In all manifestations of hierarchy, whether in the ranking of the elders' procession into the service or in the prescribed order of revealing visions, all women preceded men of whatever rank as a sign of their subordinate status. The chief rationale for the secondary status of women in the church has to do with spiritual power. Through their reproductive role, women are considered potentially ritually impure: the polluting blood of menstruation and childbirth, in indigenous thinking as well as that of C&S, vitiates spiritual vitality, and thus subverts the ritual process.[15] Just as women were excluded from some Yoruba *orisa* shrines, so no female member of the London C&S, however senior, or of whatever age, could set foot on the altar dais during ritual proceedings. Particular prayers from the floor would be assigned to women, but, in the 1970s, they could neither lead the service nor preach. When menstruating, women must either stay at home, or stand in the passageway outside the church hall, avoiding contact with their Bible, prayer gown, or other spiritually charged objects. All women, whether in prayer gowns or not, have to cover their heads during prayer. Apart from allegedly distracting the brothers from worship, women's hair also has associations with feminine power. Just as Samson's locks were the repository of his masculine physical and spiritual strength, so unrestrained female hair has the potential to desecrate a sacred space. As Saint Paul enjoined, it should be contained.[16]

At the start of the C&S in London, women were also excluded from the central administrative organization of the church. There were no female officers in the church administration or the Executive committee. But by 1971, they were included in both. These developments were part of a democratization that increased the involvement of members in the running of the church and tightened their relationship with the leadership. The Executive expanded, became subject to election, and brought in representatives of working committees set up to deal with aspects of church life. All these bodies came to include elected representatives from the Bands. Although many posts were somewhat nominal, and meetings often poorly attended, the proliferation of administrative roles was out of all proportion to the rate of congregational growth. By 1971 there were 283 offices to be held, besides the plethora of posts in ad hoc committees, and extra roles invented by every group for this purpose or that. The year 1970 also saw the introduction of an open Annual General Meeting (AGM). During a meeting in 1969 the leader explained his ambitions for the London C&S: the members, rather than the Executive, should "do the job of the church," thus "ruling people by themselves." "This is a new experiment for the smooth running of the church," he said; ". . . we are in England now, so we must have a different organisation to suit this country. . . . When we go back to our own country we will all be leaders. We are not today, but tomorrow, as qualified people, we will be in high-up positions. So why should we not be our own leaders in our church?"

Democratization: Secular or Sacred Power?

This evolution of C&S organization toward representation and delegation of authority did not always proceed smoothly. During wranglings at meetings, the language used was sometimes more that of the political party or the boardroom than the vestry: the need for officers to "report back to the electorate," the "cheated shareholders," of a badly run "company." Countering such language in one troubled meeting, the leader attempted to assert the spiritual purpose of the church: "If this was a firm then we will have shareholders, but this is a different thing, we are doing a religious organisation . . ." (First AGM, November 1971).

The contrast between matters of organizational and spiritual efficacy surfaced continually. The evolution of church organization, the shedding of ritual prescription from C&S constitutions, the search for appropriate structural forms, the increase in administrative positions, the role of structure as a site for personal ambitions, the use of market models to shape the language of debate—all do indeed speak of a new preoccupation with secular concerns.

With received Weberian wisdom, we well might interpret this intrusion of organization into the spiritual discourse of the church as the swamping of the sacred by the secular. In the judgment of Bryan Wilson, in his *Magic and the Millennium* (1973), the lesson to be learned from the history of C&S is that: "Organisational stability and rational procedures stand in sharp contrast with thaumaturgical pursuits, which congregational participation does not alone resolve" (172). Wilson fully grasps that thaumaturgy, the "central idea of gaining spiritual power" (164) motivates C&S as it does many other movements. The problem, as he sees it, is for "West African independent churches . . . to retain . . . the thaumaturgical operation, but to diffuse and democratize its practice, whilst establishing a hierarchy of control, which in itself partly rests on distinctions of graduated thaumaturgical power. This is both an institutionalisation and in some measure a rationalization of thaumaturgy, and its long-run consequences may well be the increasing subjection of the thaumaturgical to purely organizational goals" (170).

But Wilson also argues that: "For a time these instruments [new constitutions and new structures] facilitate religious association— perhaps for a long time" before they undermine the search for healing power (497–498). As Evangelist Olatunji argued during the fraught 1971 AGM, any association needs a "stable administrative organisation, without this, it can never stand." What was needed in the U.K. C&S, as in the Northern Movement, was an organization appropriate for its particular congregations, men and women whose class and life experience demanded a centralized but democratic structure for the efficient management of the power of the Holy Spirit. The style of longhaired Bar Beach prophets in Lagos was not appropriate for a would-be elite. Growth of bureaucracy in the church, therefore, did not necessarily entail waning interest in the operation of unseen forces. Secularization, the despiritualization of the universe, is a different matter. To this we will return (chapter 9), for it is this, I shall argue, not organizational forms, that effects "rationalization." When spiritual power ceases to be significant in the construction of meaning, then the demands on the church must change. It was not structure that would push the Spirit out of Aladura; only if the Spirit departed from the life experience of members would the C&S follow the path of the major Christian denominations in shedding charisma.

For the Yoruba community in Britain in the 1970s, as for previous generations in Nigeria, the C&S offered a source of divine efficacy for those who found this concept relevant, whether as a consistent etiology, or as a potential response to particular circumstance. Accommodating to the diaspora context, the elders strove to construct an organizational form appropriate to the social composition of its constituency, to appeal to "Stars of the World." This adjustment amounted to adaption and development rather than radical innovation; the basic principles of C&S organization remained constant, as did the fundamental tenets of an Aladura belief. Guided by its divinely legitimated history, yet unshackled by rigid structures, the church had the capacity to adapt to its members' needs. The organizational trajectory of the U.K. C&S ran from Saint Moses Orimolade, through the Northern Movement, to the church in Britain. From Lagos to London it retained its character as a Yoruba institution consistent in its ritual objectives: the generation and deployment of spiritual power.

This page intentionally left blank

Chapter 4

From *Aṣẹ* to *Agbara*: The Concept of Spiritual Power in the Cherubim and Seraphim

If the C&S church was merely an association aiming to provide a network of mutual support for migrant Yoruba communities, its role could have been fulfilled by a social organization or ethnic union. But the efficacy of the C&S for its members lay in its application of Yoruba cosmological principles in the management of life events. By investing new situations with familiar meanings it enabled members not only to make sense of their present but to have a handle on their future. The key concept in this epistemology—and therefore of the religious praxis of the church—was that of spiritual power: the unseen principle of efficacy that emanates from God and articulates his creation through a web of energy. Subsequent chapters examine power at work, as the basis for ritual action—for prayer, divination, and possession by the Holy Spirit. But here it is power in Aladura discourse that concerns us; what the Cherubim and Seraphim *say* about power, whether implicitly in allusion and imagery, or in conscious explanation.

The Search for Spiritual Power

There is Power, there is Power,
There is Power in the Blood of Jesus.
Send down the Power, send down the Power,
Send down the Power, O Lord.

(Popular choruses in C&S services)

This church is based on spiritual power. Without spiritual power there would be no church.

(Brother Fisayo)

The quest for spiritual power lay at the heart of indigenous ritual; the same key organizing principle was carried forward into Aladura practice.[1] The C&S pray for power; they sing about it, dance for it, and embody power in possession. Power is the subject of dream and vision, and the acquisition or threatened loss of power is the sanction for morality. Ritual techniques seek to control spiritual energies and deploy them for practical ends.

To claim that spiritual power is a central concern for Aladura, as for the majority of AICs, is nothing new. Analysts of the movement have argued for the significance of the search for thaumaturgy, that is, "the belief in, and the demand for, supernatural communications and manifestations of power that have immediate personal significance" (Wilson 1973: 70).[2] But what has claimed less attention in the literature on African sects is an exploration of how those members engrossed in the search for spiritual power conceptualize the object of their religious endeavor. Far from being considered as chaotic energy running amok in an insecure universe, the efficacy of the concept of spiritual power lies in its being mysterious but familiar, elusive but predictable, volatile but controllable. Just as the C&S themselves justify their explanations about power by appealing to perceived regularities in their own experience, so it is possible to deduce from diverse discursive practices the principles upon which this elusive essence is believed to operate.

The majority of members of independent churches would not necessarily formalize these principles in words; spiritual power is sought and manipulated, but not always analyzed. Just as only a minority of individuals in Western societies could outline even the basic principles of electricity on which their daily life depends, so spiritual power for many Aladura remains a given. What they are looking for is results through effective prayer and ritual, rather than power per se. In one way this disregard is salutary; focus in analysis should not overdetermine saliency in emic discourse. But there was also a marked—and perhaps unusual—discursiveness on the subject in the C&S itself. Whereas in indigenous prayer and divination, spiritual power is assumed, in C&S discourse it is continuously made explicit. Members were constantly reminded of the principle behind Aladura ritual: power was talked about, whether spontaneously in sermons and Sunday schools, or in reflective discussion following my enquiries. Revelation, messages from God derived from trance-inspired visions, also offers rich insight into Aladura thinking. Vision texts communicate both the ubiquity and the character of the Holy Spirit's power.

It may be that the necessity to mark their distinctiveness, first in the mission context in Nigeria, and then in the secular social environment in London, provoked a degree of theological self-consciousness on the part of members. They also—unlike contemporary New Age seekers after spiritual power—had an indigenous blueprint on which to build. Church leaders, especially those who were most concerned with spiritual matters, offered concise accounts of the nature of power and its significance in the ritual life of the church and the personal experience of members. It was this expertise in its ritual application, acquired over time, that constituted the symbolic capital indicated by their grades in the C&S hierarchy. In what follows, my construction of the concept of power is grounded in their own.

What emerged from these discussions and the analysis of Aladura texts was that although the C&S insisted on their Christian idiom, they shared Yoruba cultural assumptions about the nature of spiritual power. In spite of their efforts to distinguish C&S practice from indigenous ritual life, it became evident that they had not divorced themselves from the fundamentals of Yoruba discourse. Rather, this was what informed their own reading of the Bible and their evaluation of both orthodox and Aladura Christianity.

Christianity, "Paganism," and Power

That it was the concept of power itself that provided this continuity between tradition, Christianity, and the C&S escaped early opponents of Aladura. Although profoundly concerned with indigenous superstition, missionaries and early European observers do not appear to have pondered deeply on the logic of the cosmology they encountered and the concepts that articulated the great diversity of Yoruba ritual practice.[3] Christian Yoruba scholars of the old school were more concerned with manifesting the coherence and sophistication of the indigenous Yoruba theological system than with analyzing the principles behind the ritual process. Johnson's 1921 *The History of the Yorubas*, Fadipe's *The Sociology of the Yoruba* (1970), and Idowu's classic *Olódùmarè: God in Yoruba Belief* (1962) did not explore elements in this cosmology that were inimical to mainstream Christianity.

It was not until 1966 that Pierre Verger directed attention to the Yoruba equivalent of this *mana*-like "vital power," and identified it as *aṣẹ*. He said that *aṣẹ* is "power itself in an absolute sense," of which "divine powers" are only particular manifestations. "Nor are the Gods the only beings animated by it: it is the principle of all that lives or acts or moves. All life is *aṣẹ*. So is everything which exhibits power, whether in action, or in the winds and drifting clouds, or in passive resistance like that of the boulders lying by the wayside" (1966: 36–37).

It is impossible to encompass *aṣẹ* adequately in translation. But recent works that explore semantics and discursive practice in order to illuminate aspects of contemporary Yoruba life have had to grapple with this central concept, or, as Drewal has it, this "Yoruba theory of action" (1992: xix), for the meaning of *aṣẹ* emerges most clearly in the context of specific activity. They all have endorsed Verger's conclusions, emphasizing those aspects of *aṣẹ* relevant to their particular analyses. Gleason, absorbed in the conjuncture of physical and psychic worlds, glosses *aṣẹ* as "a sacred power, an active essence, which flows in varying intensities through plants . . ., through animals, people, and the Orisha" (1987: 300). Apter (1992: 84) stresses the political: *aṣẹ* is the "mystical force" derived from the *oriṣa* (gods), which is manifested in the "sacred power and authority" of the king. It is "ritual power, command, authority, efficacious speech" (243), as well as "the capacity to invoke powers, appropriate fundamental essences, and influence the future" (117). Margaret Drewal is also concerned with the political and social: "when Yoruba speak of an individual with *aṣẹ*, *alaṣẹ*, a person with authority, they usually mean one with innate metaphysical power who by virtue of this power maintains complete and awesome control over spiritual realms and, by extension, over social ones" (1989: 203). In her study of Yoruba ritual "play," *aṣẹ* is "performative power; the power of accomplishment; the power to get things done: to make things happen" (1992: 201). *Aṣẹ* "acknowledges innate individual power and potential" (Drewal and Drewal 1990: 6). As one *oniṣegun* (a master of indigenous medicine) put it to Hallen and Sodipo, ". . . people are different from each other and some have more power *(aṣẹ)* than others. Their *ẹmi* [spirit] is different from others" (1986: 111). In the world at large *aṣẹ* is this "power and potential" present in all things, animate and inanimate, secular and sacred (Drewal 1989: 203); in short, the Drewals argue, "The fundamental concept of life-force—that exists in many forms and manifestations and in varying amounts—is at the foundation of Yoruba philosophy and social organisation." (Drewal and Drewal 1990: 6).

Power as a central organizing principle; power as process, action, relationship, as recognized authority, as the source of skills and personal energies, as the potential to influence the future, as the secular and ritual energy that articulates the universe—these are the threads of continuity that can be identified between indigenous discourse and that of Aladura. How is it then that when discussing power with C&S, they flatly deny that *aṣẹ* is any such thing? Power, they maintain, whether of God, of invisible entities, or as manifested in individual aptitudes, is *agbara*. *Agbara* is the word that runs through C&S

prayer, preaching, and revelation. Whilst the *onise gun* who discussed *aje* (witchcraft), with Hallen and Sodipo (1986: 111) refer to this energy as *ase*, C&S elders gloss the power of witches' *emi* as *agbara*. In Aladura terminology, the force animating enemies is *agbara emi Esu*, the power of Satan, and the ritual power and "earthly" skills achieved through the C&S are aspects of *agbara Emi Mimo*, the power of the Holy Spirit. *Agbara* is glossed in Abraham's Yoruba dictionary (1958: 27) as "strength," and was once essentially a physical concept; the word also refers to sexual prowess. In terms of nonhuman force, *agbara* does appear to have been used in relation to medicines, but the innate efficacy of medicine, as of everything else, was referred to as *ase*.[4]

Ase has not slipped out of the Aladura vocabulary altogether: it appears in the restricted sense given by Abraham (1958: 71) as "authority" or "command." This is reflected in the common congregational conclusion to prayer: *ase*, a traditional ritual response that indicates more certainty—"it will be so"—than the supplicatory Christian "let it be so" represented by *amen*. The term also appears in the text of prayers:

> Before the sun could be created, before the moon could rise, it needed your authority, your command.
>
> (Translation)

The Yoruba word used here is *ase*, but refers to God's ability to issue effective orders rather than his generative spiritual power. Given that the sanctions endorsing these commands are supranatural, something of the old *ase* lingers in this authority; the word is not divested of spiritual meaning. But according to experienced C&S members it refers to one particular aspect of spiritual force rather than encompassing the whole: "*Ase* is the power of effective utterance," (i.e., the performative power of words), said one. "*Ase* is one type of *agbara*, one kind of spiritual gift. So if you have the power of effective prayer it is *ase agbara adura*," said another. "Some people have natural *ase*; whatever they say must come to pass." Rather than *agbara* being a manifestation of *ase*, it now seems as though *ase* is but one aspect of *agbara*.[5]

Why has Aladura effected such a transformation? The answer seems to lie in the Yoruba translation of the Bible sanctioned by the Anglican CMS. Of the scores of references to power in the Old and New Testaments, nearly all, whatever their semantic shading, are rendered as *agbara*. *Ase* seldom appears other than as translating the word "seal," or referring to the commanding authority of God.[6] *Agbara* stands for the "strength" of Samson[7] and for the Lord's "strength of his arm,"[8] but the same term is used for the "power" of God when this clearly embraces a variety of energies.[9] Thus the superior power of God praised by the psalmist, and his capacity to crush opponents, is all *agbara*,[10] although in an Aladura reading, such texts would clearly communicate more than political authority and physical vigor. But then the *ọta* (enemies) of the Psalms, the "powers of darkness" (*agbara okunkun*, Luke 22:53), together with Satan and his agents— witches, sorcerers, bad spirits—are all empowered by *agbara*.[11] Similarly, *agbara* stands for the power of Jesus and his disciples over "unclean spirits" and disease,[12] evil,[13] and life and death itself.[14] The power of God over nature and the elements is translated as *agbara*,[15] as even, on occasions, is the power inherent in words.[16] Visioning and other spiritual gifts as well as more mundane skills and talents are demonstrations of *agbara Olọrun*, "the power of God."[17] Miraculous power is sometimes rendered as *ẹmi* as well as *agbara*;[18] passages that are critical for the communication of the power of the Holy Ghost, such as the immaculate conception,[19] the Second Coming,[20] and the promise and fulfillment of the Day of Pentecost speak of *agbara Ẹmi Mimọ* (power of the Holy Spirit).[21] But when the C&S proclaim the crux of their practice in the Lord's Prayer: "For thine is the kingdom, and the power, and the glory . . . ," power is simply *agbara*.[22]

Thus translated, the impact of spiritual power in orthodox Christian discourse, as communicated through its central text, is heavily reduced. The vital principle of *aṣẹ* sits awkwardly in mainstream Christianity, which emphasizes morality more than miracles. The ineffability of spiritual energy is exchanged for the prosaic concept of divine omnipotence, confirming God's majesty, not his magic.

This is a far cry from the swirling energies of the *orisa* Oya,[23] the vital force made visible in *Gelede* masquerades,[24] the flux of male and female potencies of *aṣẹ*. And this, most probably, was precisely the intent. The Yoruba Bishop Crowther, coordinating the translation of the Bible in the 1870s and 1880s, had little truck with indigenous metaphysics. His dictionary (1852: 47) glosses *aṣẹ* as "a coming to pass; law, commandment; virtue; effect; imprecation," and *agbara* as "strength, might, ability" (13). How far this represented a personal ontology divested of spiritual vitality it is hard to say; Crowther's attitude to indigenous religion was ambivalent.[25] But certainly *aṣẹ* was too heavily implicated in the indigenous cosmology, bore too much symbolic weight, to be incorporated into the new religion. Christianity had to be presented as a radical break, a move from "darkness into light." The shift from *aṣẹ* to *agbara* is part of this agenda.

We have little evidence on how much consensus once existed around the understanding of *aṣẹ*, and certainly cannot draw conclusions from contemporary explanations: some informants in London disavowed this vernacular terminology for spiritual vitality or rejected the concept of an all-pervading power altogether. Certainly the nuances of *agbara* are now contested. Although the assumptions of the Yoruba spiritual economy can be read back into Aladura *agbara*, the change of terminology had significant implications. By divesting the term for spiritual vitality of *aṣẹ*'s semantic load, the way was opened for greater flexibility in the interpretation of the whole concept of power by the C&S, either as more instrumental and less immanent, or as more nearly approaching the remnants of *aṣẹ* in the orthodox notion of sacramental grace.[26] The use of *agbara* allows for a wide range of meaning, from indigenous notions to those of mainstream Christianity, relevant for an independent church in changing circumstances.

Since much of the C&S ritual proceedings in London were conducted in English, as were my discussions with members, the actual word "power" was much in use. Although carrying intrinsic dimensions of unseen vitality, the term in an Aladura context encapsulates many of the connotations of English usage, which commonly concentrates on force and political authority. In popular use, power has politico-military significance as the ability to impose policies—or policing—on an unwilling public. These meanings are indeed packed into the secular Yoruba use of the word, and the unseen world of witchcraft and medicine, as we shall see, is precisely about the attempt to persuade or bludgeon others to bend to the user's will through the exercise of spiritual energies.

However, in Yoruba semantics, the negative baggage carried by the term is much less weighty. Power is a ubiquitous part of Western Nigerian vocabulary, and in popular discourse the promise of power has been used to advertise anything from patent medicines to aspiring politicians: "Aspro for Power." Guiness promises "Power and Goodness"; Chief Femi-Kayode canvassed as "Femi-Power" in the early 1960s. If understood as "the ability of individuals, or members of a group, to achieve aims or further the interests they hold" (Giddens 1989: 729), then power does not necessarily imply conflict. In this broader sense, power becomes a dimension of action, as the capacity to control those resources necessary to alter a course of events and produce a desired result. Thus, as Foucault has argued (1980: 119), power may be creative, constructive, and pleasurable, rather than morally repressive. In the Yoruba context, partly because of the slippage between secular efficacy and spiritual strength, a powerful position, or the manifestation of personal power in successful and purposeful action, provokes admiration even when exercised at others' expense. Power predominantly signifies physical, social, and metaphysical strength, which commands compliance and respect. This recognition both constitutes and augments the status of the powerful. A universal characteristic of power is that it implies relationship;[27] power is thus inherent in all action as effective agency, which impacts upon others or on the environment. This (inter)active quality of power is particularly critical in Yoruba thought; just as power loses much of its negative value in Yoruba discourse, so it sheds something of the static quality acquired by the term when it is primarily associated, as so often in English use, with formal rank and position. To Yoruba, power implies process, movement; it is the opposite of stasis and passivity.

It is necessary to consider the concept of power in Yoruba discourse at large, for indigenous assumptions on patriarchal, political, and socioeconomic authority shape the meanings associated with power in its sacred dimension. This is not just because secular relations provide a template for spiritual discourse, although, as we shall see, sociological realities are indeed reflected in the significance accorded to the acquisition of power. It is rather because power carries a multitude of meanings, which blur the distinction between metaphysics and materiality. The contrasts maintained in Western dualistic philosophy between body and mind, matter and spirit were foreign to Yoruba indigenous thought, and are blurred in the way that C&S understand their world. I once asked an elder—a structural engineer—whether there was an indigenous concept of biomedicine. "Of course," he replied. "If you take the bark of a certain tree and give it to a witch to drink, she will die." In popular Yoruba discourse, secular and spiritual still elide, with activity in the unseen sphere affecting visible relations, and quotidian events impacting upon hidden forces.[28]

Power in Aladura Discourse

The theme of power runs through all C&S supplication, but the different aspects of this spiritual efficacy emerged most clearly from prayers specifically requesting the presence of the Holy Spirit—the source and vehicle of divine efficacy. What follows is the text of one such prayer for consecration, that is, for empowerment through the blessing of the Holy Spirit on the congregation attending a Watchnight service, a late-night gathering focused on the revelation of messages from God through vision. The translation (as of the Yoruba prayers and revelations cited below) is that provided by a prophet of the C&S. In the original, the word given as "power" was *agbara*.

O Lord God Almighty, Father, Son and Holy Spirit, we thy humble children are before thee at this hour of the day. We ask for consecration from you. O Lord God of the Soldiers,[29] come and consecrate us. Our service of today, come and consecrate it. The four corners of the house, come and consecrate
5 it. Our song, those that we are about to sing, our psalms, the ones we've said and the ones we are going to read, come and consecrate it. All the people kneeling down before thee today, come and consecrate them. Those that are coming on their way, come and consecrate them. Those that can't come, wherever they may be, come and consecrate them.
10 We ask for thy Hosts of Heaven, those that are strong in power and sense [*a beere fun awon omo ogun orun re to po ni agbara ati ipa ati oye*]; those that give people wisdom; those that change life for the better; those that cast away obstacles in people's way; those that help you so that you have full opportunity. We are asking for thy Holy Spirit [*Emi Mimo*] by which we will be able to
15 achieve our aims. Thy Holy Spirit which it pleased you to give to Moses in those days, that he passed through the Red Sea, and by which you gave him victory over enemies [*ota*]; this type of Holy Spirit, come and descend it down. The Holy Spirit which you gave to Daniel in the lion's den, so that we, young and old, will be able to conquer the enemies, like aged and female members
20 [i.e.,witches], this type of Holy Spirit come and endow us with it. The type of Holy Spirit descended to those three men thrown in the furnace, Meshach, Shadrach and Abednego, and the fire didn't burn them; O Lord, send it down. How are we going to prove that we are C&S here and in Heaven, except you give us life [*iye*], except if you give us your Holy Spirit, except our obstacles are casted
25 away?
We beg and ask thee, God of Hosts, O God of Life [*Elemi*], God of Words, God of Light, come and endow us with thy Holy Spirit. Power of pure vision, power of deep ones, and power of true ones, come and endow us with it. The power of love, that the devil will never deceive us, that the enemies
30 will not overpower us, that the church will be growing in strength and that we will respect ourselves [i.e., each other], come and endow us with it. The type of Spirit that you descended on Samuel since his youth—he performed wonders with that, and miracles. The type of Holy Spirit which you endowed Elijah with; the same that you endowed the Apostles of old with. They spoke
35 in tongues, and interpreted. The power of true vision, the power of promising vision [i.e., prophetic], the power of deep dreams, and power of deep visions; the power of speaking in tongues, and the power of interpretation.
Jah Jehovah! Jah Jehovah! Jah Jehovah! The Holy Spirit in Heaven come and descend it upon us. The power that we will never go ashamed, that

40 our prayer will never go unanswered, that the pregnant women will bear easily, that you will provide for those looking for children; people looking for job
will get good jobs, that the fold will be multiplying. The power by which we
shall be prophesying, and all our prayer will be accepted by thee. O God of
the Holy Spirit! Lord God of the Hosts of Heaven! The power for joy, for
45 rest of mind, that we will never have trouble in our homes; the power to
fight all the impediments in our ways, that we will never be ashamed. We
beg thee to come and endow us with thy power. All the disturbances and
unrest of mind, come and cast them away from us, so that the service of
today might be accepted.
    Through Jesus Christ our Lord, Amen.

(Elder Oterunbi 9 January 1971;
translation)

From this prayer, as from all C&S ritual texts, it is clear that the
Aladura concept of power carries over meanings from popular usage.
It is more than efficiency or good fortune; it is essential for a produc-
tive existence because it is associated with life itself.

Olodumare (the Almighty God), who is also addressed in prayer as
Olọrun (God of Heaven) is the ultimate source of this vitality in both
indigenous and Aladura cosmologies. For Aladura, he is *Jah Jehovah*,
God of Power (line 38). He is also God of the Holy Spirit (line 43).
Indeed, (spiritual) power and the Holy Spirit are used interchangeably
by C&S in their English translation of *Emi (Mimọ)* (line 14). He is also
addressed as *Elẹdaa*, the Creator. He is *Olọrun imọlẹ*, God of Light
(line 27), and so (as we shall see in chapter 5), connected to spiritual
power. He is often worshipped as *Olọrun Iye ati Agbara*, the God of
Life and Power. The intimate connection between life/spirit/power
apparent in lines 23–25 of the prayer is embedded in Yoruba semantics.
The Yoruba for "life" is most often rendered as *iye*, "the quality most
associated with power" (Peel 1968: 136); the word is constantly called
out as a spontaneous prayer during services. But in the appellation
Elẹmi, Owner of Spirit or Life (line 26), the concept of life is linked
with *emi*, spirit. *Mi* means breath: that which differentiates the living
from the lifeless; implicit in the references to *emi* in the consecration
prayer is power that generates breath. According to a C&S prophet,
"anything which is breathing is *elẹmi*, it means to be alive. But spiritually it means that which Holy or bad spirit is using, one which has spirit
in any sense." Abrahams (1958: 187) glosses *emi* as "life," and in the
discourse of traditional ritual adepts *emi* could refer to the person's
"spiritual component," which "can expand into various dimensions of
special abilities, talents, or powers" (Hallen and Sodipo 1986:
114–115). The invisible, enduring aspect of an individual is also rendered in Yoruba as *ọkan* and in English as "soul"; Prophet Korode

translated a recording of one of his prayers thus:

> O God, you who own the soul (*ẹmi*) of life (*iye*), your spirit (*ẹmi*), come
> and descend it on us.
>
> (Service 5 September 1970)

On another occasion he tried to illuminate the concept of spirit by
saying, "*Ẹmi* which is *agbara* is *iye*; once your soul leaves you, your
body's empty, so they are all like the same thing."

When C&S elders struggle to convey these interrelated meanings,
they use illustrations with both indigenous and Aladura referents, for
the discourse on power is shared:

> Power comes from life, life from breath, and from your spirit comes the
> power. Yoruba say, for example, "We believe that the spirit of your dead
> grandfather still exists, and that you can invoke his spirit, invoke his
> spiritual power."
>
> (Pastor Tomori)

Or again, in reference to powerful entities:

> The strength they have is their power, and the spirit is where the power
> is coming from.
>
> (Prophet Korode)

The notion of spirit, then, as activated by power, is a dynamic
one, a dimension of the central concept of spiritual power. When
taken together in C&S discourse, it is clear that spiritual power is
the energy that quickens matter. C&S prayers, as in lines 22–25
of that quoted earlier, imply that weak *ẹmi* entails general
enfeeblement:

> Descend thy Holy Spirit, O Lord, thy Holy Spirit without which we can-
> not do anything, without which we are dead like a yam which is rotten
> inside.
>
> (Prayer. Service 7 March 1971; translation)

> And now we ask you for Your Holy Spirit, because without it our life is
> empty and we are nothing.
>
> (Prayer. Service 6 November 1969; translation)

Visions communicate the same message:

> And to our Brother Oduntan. The Lord says that you must pray fervently
> against anything which may make you lose your spiritual power . . . . I do
> not want your life to be like a dead fish, like an old shoe . . .
> (Vision. Elder Sister Adeoye.
> Service 20 April 1969)

Power is a principle of general efficacy. Existence without the life
force of spiritual power is worse than useless, for it has no purchase on
the world. Spiritual powers enhance personal influence, and success
is the product of spiritual strength. Since, in the Yoruba cosmology,
there is slippage between the spheres of *soma* and spirit, successful
activity in one predicates the vitality of the other. Strength is at once
both the source and the evidence of an effective persona, whether
speaking spiritually or in a secular idiom.

The concept of power as lying behind a successful life has its spiritual counterpart in C&S cosmology; any manifestation of a successful life can be seen as a refraction of the Spirit. The consecration
prayer does not merely list problematic areas of life where God's
help is needed, it asks for power to deal with each eventuality; petition about practicalities is phrased in terms of empowerment,
the "Holy Spirit by which we will be able to achieve our aims"
(lines 14–15):

> We are in a foreign land; we are sojourners in a foreign place. So we are
> asking for thy power. Thy power is wisdom, thy power is knowledge . . .
> (Prayer. Service 7 March 1971)

The acquisition of spiritual power to combat the fragility felt by
students in an alien environment was also recommended by Prophet
Korode in a Sunday School lecture (8 June 1969):

> Instead of saying "O God, let me pass my exams," just ask for the Holy
> Spirit straight, because if you get the Holy Spirit you will understand
> anything you are learning. You will get the favour of the teacher; you
> will even be shown questions in dream . . .

The same theme runs through prayers for spiritual power, "which will
be with us in our examination halls; the spiritual power which will
enable us to remember all what we have studied."[30]

As for exams, so for other aspects of a worker-student's life: the C&S pray for "the power of administrative work," "the power by which those who are looking for provision of job will get good job," "thy Holy Spirit by which we will be having grace and favour from all men" when up against employers, examiners, immigration authorities, or the police.

The angels, the Hosts of Heaven (lines 10) may be instrumental in dispensing this power, since they are efficacious and discerning delegates of God. As in the consecration prayer they may represent particular aspects of his energy, particular fields of competence. This text is characteristic of C&S prayers and visions where the relevant type of power is sought to deal with problems of health, fertility, employment, housing, and other daily concerns. Conception and childbirth appear to be especially susceptible to the enabling power of the Spirit:

> The Lord says that we must pray for pregnant women in the fold, especially for one new member, so that she will be able to receive the type of Holy Spirit with which she will be able to deliver safely.
>
> (Vision. Sister Opayinka. Service 27 July 1969)

> The spiritual power which will enable the barren women to conceive and bear easily; that our students may be successful in their undertakings, that you will be answering everyone according to his or her wishes—we beg thee, Lord God Almighty, that this type of Holy Spirit may continue to multiply in our midst.
>
> (Prayer. Service 18 January 1970; translation)

These quotidian "undertakings" of members' lives feature in prayers for spiritual power as prominently as requests for the acquisition of spiritual gifts and proficiency in vision, prayer, and prophecy (lines 26–29, 34–43).[31] But the purpose of prayer and prophecy is to direct daily life toward "full opportunity" (lines 13–14), and spiritual power is the driving force behind all aptitudes including effective prayer itself ("the power . . . that our prayer will never go unanswered" [lines 39–40]). There is no tension between spiritual and more mundane petition. Psychological well-being is also part of the package of power. Anxiety and depression both destroy personal equilibrium and hinder ritual concentration. So disturbances and unrest of mind (line 47) must be cast away by another manifestation of the Spirit: the power for joy (line 44), with which to lead an effective life.

The Powers of Evil

The power sought through C&S ritual is a proactive force, a power *to*, generating capacities and capabilities of mind, body, and spirit. But it is also a power *over*, a protective vitality against the enemies (*ọta*) (line 17) and evil spirits (*ẹmi buruku*) who will sap this energy, and cause calamity and failure.

Who are these enemies, these evildoers, who pose such a threat to members' well-being? How do they carry out their devastating assaults?

The Lord says there are some people who are waging war against my people, who are using secret means, I mean juju, towards my people. Those who are doing that, I will send the angel of the Lord to strike you, and the people will say *Amen* [congregational response]. To witches and wizards; your meeting yesterday towards [i.e., against] the Church was in vain. And to those witches and wizards in the Church, you shall be punished, and the people will say *Amen*. And to the evil doers, the Lord said unto me, you shall be with one trouble and another, and the people will say *Amen*. The Holy Ghost is given to each and every one of us today, and may the evil spirit never take it away again.

(Vision. Brother Kolarin. Service 18 January 1970)

Indigenous medicine or juju, witchcraft and sorcery, evil spirits, together with Olodumare's counteracting power—the whole panoply of the Yoruba invisible world is brought into the heart of C&S christianity. As Ray (1993: 267–268) rightly argues: "While rejecting most of the 'pagan' belief content of the religious environment, the founders of the Aladura churches retained two fundamental elements: the belief in invisible forces, especially malevolent spiritual powers, and the belief in the efficacy of ritual action." When pressed on this point, elders in London did not deny that the entities in the indigenous religious complex are powerful; they were not negated, but denounced. Rituals for the *orịsa*, the complex range of deities under the High God, are certainly condemned as "idol-worship," and an undesirable aspect of a particular god, such as smallpox (associated with Sopona), may be characterized as an evil spirit. But it is more juju, witchcraft, and evil spirits that come in for the full weight of Aladura censure. Here we have spirit in its sense of an invisible being, together with the vitality that enlivens it. The fusion of these meanings in the word *ẹmi* emphasizes the significance of the intrinsic power through which spirits operate, and by which they are both defined. It is thus by power that they will be defeated:

> Jehovah Emmanuel, may bad spirit (*ẹmi buruku*) not be in our midst. The seal (*aṣẹ*)[32] by which the world and heaven is created, for God to send it down . . . . The spirit of sickness (*ẹmi aisan*), may it not dwell in our midst; the witch and witchcraft who spoil our work in the night, come and blindfold them . . . . Jah Jehovah [powerful God] you that called Moses Orimolade Tunolase to do thy work, the power with which you have worked with him, come and descend it upon us.
>
> (Prayer. Service 15 June 1969; translation)

Dream and vision serve to flesh out the picture of these "powers of darkness" according to traditional ideas:

> I saw a certain yellow dog going all about together with a certain band of enemies coming before the fold. The Lord says that today, the band has been destroyed . . .
>
> (Vision. Teacher Aiyegbusi. Service 4 January 1970)

> There is a lady with a gourd on her head, and this is supposed to be Holy Water, but instead it is filled with blood. Any lady trying to deceive people by telling them that she is helping them, but instead doing the opposite, should be warned, so that the gourd will not break on her head.
>
> (Vision. Prophet Korode. Service 7 November 1970)

These revelations are not merely rhetorical; they are taken as significant warnings. At the end of the service, the Spiritual Leader repeated the warning to the "lady with a gourd":

> The Lord has said it, not Korode. If you are praying for someone, but rather to give them the blessing, you are collecting his or her blood, that is very bad indeed . . .

Even in such an announcement it was not necessary to spell out the meaning of the message; everyone would have known that these visions refer to witchcraft, for the image of the witch (*ajẹ*) in C&S discourse mirrors indigenous concepts.[33] The congregation are well aware that witches, usually women, are thought to form themselves into bands, who use their powers to sap the *ọkan* (heart or soul), or the *ẹmi* (spirit, life-force) of their victim, resulting in wasting sickness, mental disturbance, or sudden death. Blood, as associated with life and reproduction, is thought to possess inherent "numinous energy";[34] by drinking blood, witches are said to prolong their own life, and increase their personal power, whilst sapping their victim's physical and spiritual strength. A gourd is both a household container and a ritual receptacle for holding powerful paraphernalia and ingredients or objects that represent the power of a particular deity, person, or office. Witches are thus said to keep their power of witchcraft in a calabash.[35]

African witches are commonly thought to invert normal social behavior; Yoruba witches conform to the stereotype. Elders explained how they operate at night, their *emi* deserting the human sphere to invade an animal or bird. C&S revelations mentioned crocodiles and snakes, familiar from indigenous witchcraft beliefs; when a visioner saw the "maize before our Sister" being eaten by birds, the woman would understand that her spiritual nourishment and earthly success was prey to witchcraft. In defiance of the female norm, witches specialize in sabotaging reproduction, and dreams of sexual intercourse with another woman or gathering of blood are evidence of their activity. Children and pregnant women are considered to be especially vulnerable:

> I was made to see some little children in a small room . . . These are the ones the Holy Spirit has been guarding from witchcraft and the powers of evil doers . . .
> (Vision. Aladura Brother Oterunbi. Service 7 December 1969)

Yoruba have a particular fear of poisoning; members are warned in vision against smoking other people's cigarettes, accepting sweets, and "eating all about." Witches are notorious for poisoning the body and spirit of their victims:

> The elders should help all members who are present today, that whatever will make them to eat bad food while they are asleep, which may lead them to sickness or sudden death, for God to conquer.
> (Dream reported in Service 27 July 1969; translation)

Witches are also held to pass on the substance of witchcraft power to others by these means:

> I was having different dreams, and sometimes I hear bad voices. When I got fed up, I asked God, well—you are a Holy God, there is nothing

hidden from you, I know that one day you will answer my prayer. Then one day I had another dream . . . I saw Prophet Fakeye holding his rod.[36] Then he asked me to open my mouth, he put his rod in it, then I vomited this black thing out of my mouth . . . Since then I have never had all the bad speeches coming to my ears, and no more bad dreams. (Open Thanksgiving. Sister Elugbaiye.

Service 6 November 1969; translation)

Later, the sister had another dream of the prophet telling her that "what won't allow me to sleep throughout the night is what they [the members of another church she visited] have given me to eat." Thus revelation confirms familiar stereotypes of "wicked forces." As at home, C&S members consider dreams as reliable evidence of secret malevolent activity, so will be alert to anything that could be interpreted as malign attack in order to take remedial ritual action. Reluctance to mention witchcraft overtly, although partly due to a diplomatic avoidance by visioners of anything that smacks of personal accusation, is also common custom in Western Nigeria. In C&S ritual language, witches may be referred to as aṣa iye, translated by an elder as "eagles of life." By this he alluded to the snatching away of life by birds of prey—aṣa means hawk. Another euphemism is iya agba, which C&S glossed as "elderly people" or the "aged and female members" of the Watchnight prayer (line 19).

The persecution of individual women as witches was a common Yoruba practice;[37] older members in London told of seeing suspects stoned to death in their childhood villages. The C&S, however, tend to talk of witchcraft rather than witches, emphasizing power as much as personhood. But they stop short of addressing witches as "our mothers," the indigenous ritual salutation to older women's power. For in the traditional scheme of things, although witches were seen as evil people, their power could be exercised for good or harm; witches were wooed by respectful salutations and, in certain areas, placated in the masked Gẹlẹdẹ ritual.[38] The C&S recognize that witchcraft can be used as an alternative source of power: "These revelations are not from God but directly from the witchcraft," warned a vision about false prophecy (7 December 1969). But this, as all other avenues to power besides the Holy Spirit, is roundly condemned, and witches are characterized as the ultimate evil force.

Another such path to personal power is the manipulation of indigenous medicine, oogun. Oogun includes the curative use of herbs and is widely used in Nigeria for healing, protection, and to invest the user with spiritual efficacy to achieve a desired end. Oogun rere aims at producing beneficial results: success in a venture, a cure for an ailment, security from enemies. *Oogun buruku* or *buburu* is aggressive, aiming to settle scores or achieve personal goals at the expense of others. The charms and spells of medicine are popularly referred to as juju, which Yoruba in London define as to do with others rather than the self: "exerting power over another person," "influencing another's actions," usually for harm. Juju also refers, as Abidoye explained, to its components: it is "bad medicine which people use against someone, for the downfall of someone, or to set difficulties in the way of other people." Materially, juju consists of combinations of ingredients that are intrinsically powerful, or whose inner power is activated by incantation (*ofo*). Invocation may also invest material objects with the power of a deity or spirit. Words themselves may have intrinsic performative power, hence *Olorun* as the God of Words (lines 26–27). Or they can acquire this when efficacy is conferred by medicine; *epe*, "the curse," is often cited by Yoruba in London as the cause of their ills. Charms and protective amulets prepared with medicine are a ubiquitous part of Yoruba life; the characters in contemporary popular fiction and theatre resort to *onisegun* (adepts of Yoruba medicine), or *babalawo* (diviners who are often healers too), for protection, retaliation, or for their own aggressive purpose.[39]

But it is not possible to draw hard and fast distinctions between juju and witchcraft by contrasting the manipulation of materials with invisible malignity as has been done in much of the classic ethnography on witchcraft and sorcery.[40] Leighton et al. maintain that "malign influences of different origin interpenetrate each other, and the distinction between them is not clear in Yoruba minds" (1963: 113). In Yoruba ritual practice, the fields of *orisa* worship, witchcraft, medicine, and spirits are interwoven: *Onisegun* sacrifice to their deities; juju depends on *emi*, spirit, both as locus of vitality and as embodied power; witches are held to boost their powers with medicines. C&S discourse mirrors these interrelationships:

> The Lord says that the people who are practising juju in our midst, the time is up for them. And to the witchcraft, I will give you one Holy Name to call in the face of witchcraft, and they will be powerless. Hell is open for the witchcraft (*emere*) today. Anyone who tries to block the way of the Holy Spirit, they will receive the punishment today.
>
> Members should be patient to hear the voice of God; I want to reveal the secret of witchcraft (*aje*). I want to show you the ways they form their things, the mixture (*agbo*) that the witchcraft use, the bark, palm oil and so many things, and the oil they use to rub their faces to see many things in the world. If anyone should leave palm oil in a basin,

and put it in a room, in the morning-time they will find it missing or dried up . . .

(Consecutive Visions. Aladura Brother
Oterunbi. Watchnight Service
18 January 1970; translation)

Here we see the merging of the medicinal and the psychic activity of witchcraft, together with a statement on their material existence (stealing of the palm oil), which closely follows Yoruba concepts. When represented in C&S discourse, this fusion of evil identities is intensified; the prophet who translated this text for me glossed both *aje* and *emere* as "witch," even though *emere* is a separate class of spirit.[41] The exact classification of the type of enemy is less important than its unwanted impact; the actual channel of destructive power is often left unspecified.

The elision of a variety of malign agents as evildoers and "the powers of darkness," despite their different styles, is based on their common exercise of power with negative intent. The antidote against the assault of any particular baleful force may be strengthened by a specific ritual act or object, but shares with all defensive ritual the common purpose of producing a countervailing energy. What is revealed by C&S texts is the pivotal battle between the Holy Spirit, however ritually represented, and evil forces operating by whatever means.

The predominant metaphor in C&S for all this malignity is the "enemy" (*ota*), mobilized by the exercise of destructive energy.[42] C&S revelation is shot through with warnings of the malicious powers of ill-wishers working to literally overpower individuals and the church (lines 29–30) by putting "impediments in our ways" (line 46), digging pits to trap the unwary, binding victims with ropes and chains, crushing, blinding, and smothering them, enfeebling them with sickness and disease, threatening the vulnerable, and snatching success from the grasp of the fortunate.

And to our Teacher Opaleke. There is a bottle standing before you which is about to be filled with water, and this is your work in Great Britain. But you must pray that the little which remains for the bottle to be full up, that God will give you the actual water. Because I saw the enemies trying to put black ink inside the clear water. The Lord is showing you this not to discourage you, but to help you, so that success will be yours.

(Vision. Teacher Korode. Service 1 June 1969; translation)

I could see a body of water, and in this I could see a certain type of machine, of engine, going deep into this water . . . I was made to see

Brother Akinsanya also going deep into the water, and the engine was following him, and I thought he would be injured. But luckily for him, he got out uninjured. God is telling you that no matter however great your trouble may be, he is going to be with you, so you may not be overcome by any power . . .

(Vision. Elder Adejare. Watchnight
Service 5 September 1970)

To Brother Aluko. He should raise his voice three times over his family for whatever may bring bad events or happenings on the family or himself for God to cast it away. For the enemies have combined together especially in the month of July to bring disaster either on the wife, children, or himself. But from today, this has been conquered. All that the enemies have planned, their plan has been forfeited by God.

(Vision. Elder Oterunbi.
Watchnight Service
9 Janauary 1971; translation)

C&S vision also confirms that the long arm of the enemy can reach out to Britain from Nigeria:

And to our Brother. I saw a big snake before you, and this is the enemy from abroad. Anything which may make you weep in this country, call on God to cast it away . . .(Vision. Evangelist Somefun. Watchnight Service 9 January 1971; translation)

To Aladura Sister Oloko. [Seven elders must pray] against anything which can cause you sudden death . . . Don't think that there is anyone in this fold oppressing you . . . It is from the place where you were born, and the type of people there.

(Vision. Teacher Korode.
Service 18 January 1970)

Only God can forewarn of these dangers, only he can be trusted, for friends may be wolves in sheep's clothing:

Watch and pray. The enemies are digging holes in your way so that you may fall in, and these enemies are appearing as friends . . .

(Vision. Prophet Aiyegbusi.
Watchnight Service
5 September 1970; translation)

This means that despite assurances of solidarity in "the fold," fellow "brothers and sisters in Christ" are not exempt from suspicion of harming each other, and trying to disrupt the church:

> We ask and beg thee, because these people are part and parcel of us, and they are in the same fold, whoever is doing bad work, whoever could be causing some harm and finding some fault on their own brothers, O Lord God of Hosts, come and change their mind. The people who try and cause misfortune to their friends, and find ways to catch him, and the ways to destroy him, this is not the type we want in the fold. The bad people stopping the happiness of the church, hindering and stopping the prayer, come and change their minds . . .
> (Closing Prayer. Service 15 November 1970; translation)

Here, God is being asked to dissuade "bad people" from their harmful intent, but the idiom of the battle against evildoers is usually more belligerent. Enemies are revealed as "waging war"; troubles and misfortune are not only "cast away" and "averted," they are "conquered":

> The Lord says that she must have three days special prayer for the war against her by the enemies, for God to conquer . . .
> (Vision. Lady Leader Awojobi. Service 5 April 1970; translation)

> . . . whatever may be the trouble or sickness before you, for God to conquer this for us . . .
> (Vision. Teacher Korode. Service 22 February 1970)

Military metaphors abound in C&S hymns:[43]

> No witches has any power / Over member of this Band. Before Michael and the Cherub / All witches will be destroyed.
> Get ready, get all ye ready / To wage war on all witches / And all people who are wizards / And all wicked incarnate.
> Trouble and forever trouble / Will befall all the witches / Before the Army of Jesus / By the power of the Trinity.
> (Holy Hymn Book;
> ESO. Morning Star: No. 89)

Of the body of hymns that were sung regularly in services from 1969 to 1972, over one-third deal with victory over enemies with their evil strength, and the protection provided to the C&S by the power of God. If those hymns borrowed from English hymnals are excluded, the proportion is even higher. Prayers ask God for "the power to wrestle against enemies"; in the announcement of the thanksgiving donations made during Sunday services success is often summarized as "victory over the enemy," when "the Devil was waging war." While promising to empower members to lead successful lives, the C&S constantly warns them that this power will be vitiated by countervailing energies; the preoccupation with assault from others is thus continually kept alive.

External Causation

This fear of malign attack was not based only on an apprehensive imagination; there is no doubt that the practice of juju had made its way to England. In the 1980s, Lawuyi (1988: 4) found that the use of juju was ubiquitous among Christians in Nigeria as well as others. In London, apart from the confessions received by C&S prophets about the use of "medicine" (see chapter 8), members working in the Post Office often told me how they recognized the contents of badly wrapped parcels sent from Nigeria. When Kolarin revealed his vision condemning "means," the congregation responded with a fervent— and informed—*Amen*.

It is difficult to quantify the extent to which these fears of malign spiritual power were part of the cultural baggage of Yoruba students during the 1970s. "Beliefs" are badly captured by questionnaire; the strength and coherence of the discourse of negative powers would have varied not only between individuals, but between different moments in any one person's experience. However, studies carried out in the 1960s did find such concepts common amongst Nigerian students in secondary, further, and higher education—the sectors from which migrants to London would have been drawn.[44] In Britain, Anumonye (1970) found strong adherence to such notions among fulltime Nigerian university and medical students in Edinburgh, as did Stapleton among worker-students in London (1978: 32–33). The C&S is not a cult; it does not attempt to promote an alternative reality divorced from the cultural mainstream in which to enclose its members. It aims at efficacy within the wider society, building on notions of cultural truths already present amongst its potential constituency. C&S praxis could have had no resonance unless Aladura explanations accorded with existing epistemological schemas.

So when Yoruba students arrived in Britain, they did not necessarily leave behind the belief that the causes of failure and misfortune may well lie outside personal inadequacies and be attributable to the malign agency of others. In a discussion about her work with both fulltime and parttime students during this period, the Welfare Officer at the Nigerian High Commission spontaneously commented:

> Some of them have the age-old belief that people have at home. Many think that they are getting nowhere because someone doesn't like them, or bears grudge or jealousy against them, and is using poison against them, or casting spell, or using other means to cause their downfall. They think this because no matter what effort they put in they can't succeed.
>
> (Interview 28 October 1971)

I call this heuristic model "external causation," whereby difficulties and disasters are interpreted as the result of opponents working against the sufferer, using their power and influence to damage their victim. The means employed by the evildoers may be both natural and supranatural; Teacher Awofala commented, "If you don't rival one way, you rival through the other." But such a distinction is rarely made: bad faith in direct personal relationships combines with attack on the spiritual level, whilst assault can still be suspected without this being cast in a spiritual mold. Nevertheless, it is the machinations of unseen powers that especially preoccupied the C&S:

> Some people study hard, but when they take exam, all is in vain. Not because he doesn't understand or know it, but because enemies are waging war against him. You are the creator of the lot, whether devil or enemies and so on. You are the creator of every human being on earth, so all the examiners, may we find favour in their eyes; may we not have failure again . . .
>
> It is the water that quenches the power of fire, and it is the rain which quenches the dry season. Whatever may be power of enemy, conquer this one for us. Some people are having family enemy troubling them, so for God to come and conquer. We can only pray, but the answer and the power is from God.
>
> (From a Prayer for examinees. Service
> 7 November 1970; translation)

Not all problems will be explained in this way; the extent to which any one person will account for a particular mishap in terms of outside attack depends both on individual character and circumstantial context. It is especially relevant in accounting for inexplicable bad luck, in lending meaning to seemingly senseless incidents, and in illuminating the conjuncture of events in time and space, the chance factor in Evans-Pritchard's classic account of witchcraft (1937). It is one explanatory option that can be employed to inject meaning into difficulty and failure.

Sometimes you have studied from morning to night. You have done everything you can, but you still don't pass [your exam]. So then you know that it may be because someone is jealous of you and is putting impediment in your way.

(Aladura Sister Osinubi)

Say that two of you, for instance, have been friends, right from infancy. And you grow up, and you have got a nice husband and children; you have got your qualifications, and you are successful. She has got a bad husband; she has not got a home of her own—a broken home and a broken heart. Now instead of praying to God to help her out of her difficulties, if she is a witch, she will set her face against you to pull you down to the same level where she is, or to be worse off.

(Prophet Fakeye)

Cherubim and Seraphim discourse thus engages with indigenous concepts of external causation and negative powers in several ways. Primarily, it reinforces the notion that explanation of misfortune must be sought beyond personal circumstances and conduct, and may be located in the power of other individuals. This does not exonerate members from taking responsibility for their own actions; C&S ethics cast indigenous notions of good character (*iwa rere*) in a Christian mold, and full personhood (*eniyan*) is defined in terms of its requirements. Aladura also repeat the Yoruba teaching that assault of enemies may be provoked by the victim's own misguided behavior, and members were specifically exhorted to maintain good social relations to avoid retaliation. Nevertheless, as is illustrated by the visions quoted above, C&S epistemology perpetuated the explanatory framework of external causation by personalizing the factors that lie behind discomfort and dis-ease. It is on unseen dangers, hidden powers, that revelation concentrates. C&S discourse casts familiar characters into the cosmic dramas revealed in dream and vision: witches, "wicked hands," and medicine reappear in the London context, down to the details of indigenous stereotypes: the "yellow dog," the "blood-filled gourd." Visioners had to be careful how they phrased their messages, but for the oft-repeated formula "*anything* that may bring *obstacles*/sickness/sudden death" and the like, C&S congregations would read *anyone*.

The C&S thus employs an indigenous hermeneutic of spiritual powers, which operates both on existential and instrumental levels. Buttressed by revelation, external causation offers a logical explanation for apparently random misfortune, integrating malaise into a system of meaning that fends off feelings of inadequacy and lack of self-worth. Foreign perils are encompassed by the familiar; new fears are named by old labels; strange experience is reduced to the known. The Holy Spirit is then enlisted in the struggle for survival enacted through the power play embedded in both indigenous and Aladura discourse.

But, although C&S conceptions of external causation are culturally specific, they leave space for changing interpretations: the means of attack indicated by vision sometimes remains unspecified. This renders the characteristic explanation of events independent of assumptions of witchcraft and medicine, and potentially adaptable to a less spiritualized individual consciousness or shared cosmology. External causation does not necessarily disappear in a diaspora context, for even if witches and wizards fade into the background, enemies can still take center stage.[45] In the world of students in London, as charged and competitive as life in Lagos or Kaduna, C&S discourse can therefore fuel rather than assuage interpersonal tension. On the admission of Yoruba themselves, the fear of bad faith bedevils social relationships. Students in London often commented on their compatriots' tendency to be suspicious, and confirmed the significance of external causation in their interpretation of reality. Since revelation of hostility often comes through dreams, the suspect can have no opportunity for self-defence. This unilateral mistrust often exacerbates strained relationships, and appears to confirm that jealousy can outweigh support. The protective services that the C&S claim to provide thus seem all the more essential to its members.

Contested Readings: The C&S and the Bible

The C&S are thus caught in the position of officially condemning indigenous ritual praxis whilst sharing the ontology upon which it is grounded. For, despite this starting point, theirs is a Christian project; in the founding days in Nigeria they had to convince converts, if not orthodox clergy, of their theological respectability, and by the early 1970s in Britain they were actively seeking membership of the World Council of Churches. The key, now as then, in their efforts to distinguish their ritual practice from its Yoruba roots, is their dependence on the Bible as the central Christian text; the Watchnight prayer is larded with Biblical reference (lines 15–22, 32–34). But these allusions are also seen as a charter for Aladura interpretations. The Old Testament is itself informed by the conflict between monotheism and traditional deities, and so lends a historical dimension to the Aladura repudiation of indigenous practice.

But in their rejection of the *orisa*, they do not deny the reality of unseen entities. Whereas an early missionary could dismiss worshippers of Sango, the thunder god, as a "lot of fools," deluded by pagan rubbish (Stone 1900: 93), Aladura opposed the rituals they replaced as being "unchristianly," but not irrational. What they did embrace from mainstream mission Christianity was a moral dualism. The ethical ambiguity of traditional characters—whether witches or *orisa*—vanished, leaving them with negative power alone. Although C&S hold that these entities can offer alternative avenues to spiritual strength for individual benefit, they are condemned, and recast as enemies, or the means by which evil people implement their wickedness. Here C&S discourse reflects the popular Yoruba imagination. Unlike Protestant and Catholic orthodoxy, the different agents of destructive power are not denied, but are subsumed under Satan as "branches of the Devil" to be countered by the power of God. As such they are central to C&S theology:

> I think that anybody who doesn't believe there are witches and wizards can't believe there is a God.
>
> (Prophet Korode)

> If you don't believe in evil, how can you believe in good? If you don't believe in Devil and evil spirit, then you can't be a Christian. But as a church member you know that they have no effect—that you can overcome anything with prayer.
>
> (Pastor Tomori)

Yoruba worker-students living in London in the 1960s and 1970s were mainly mission-educated; indigenous cosmology was overlaid with largely Protestant orthodoxy. This worldview had little room for the operation of external malign powers or the countervailing power of God. But once Aladura converts, whether in the first or second half of the twentieth century, became engaged in the C&S project, the concept of a universe articulated by spiritual powers would have been reinvigorated rather than challenged by increasing familiarity with the Bible. For European missionaries, the gospels replaced heathen fears of evil forces with a redemptive faith in the love of God. But read in the light of traditional ontology, however unformulated and fragmented this may be, the Bible is suffused with unseen forces. Its books are laden with spiritual agency: the terrible strength of God, the miraculous power of Jesus, the exhibitions of spiritual energy by prophets and disciples, dreamers and visioners, the rival forces of witches (rendered as *ajẹ* throughout the Yoruba translation), wizards (given as *oṣo*, the common term for sorcerer), evil and familiar spirits (*ẹmi buburu* and *abokusọrọ*).

For we wrestle not against flesh and blood, but against principalities, against powers (*alagbara*), against the rulers of the darkness of this world, against spiritual wickedness (*ẹmi buburu*) in high places. Wherefore take unto you the whole armour of God, that ye may be able to withstand in the evil day . . .

(Eph. 6:12–13)

The C&S insistence that their basic beliefs are Biblical is therefore difficult to dispute. Take, for example, the Book of Psalms; this was one of the first parts of the Bible to be translated and made their own by early converts, and holds a central role in C&S ritual. Of the 150 psalms, 56 mention enemies (*ọta*) explicitly, while others convey the same message by reference to those with evil intentions who pose a variety of threats. It is striking that this theme runs through the small number of psalms that have been selected by C&S to be recited at particular points in the service, whatever the overt purpose of the prayer. So one of the main psalms for "Forgiveness of Sins," Psalm 25, reads:

O my God, I trust in thee: let me not be ashamed, let not mine enemies triumph over me.

(V. 2)

Consider mine enemies; for they are many; and they hate me with cruel hatred. O keep my soul (*ọkan*), and deliver me: . . . for I put my trust in thee.

(Vv. 19–20)

A standard psalm for Sanctification is no. 31:

Pull me out of the net that they have laid privily for me: for thou art my strength. Into thy hand I commit my spirit (*ẹmi*): thou hast redeemed me, O Lord God of truth.

(Vv. 4–5)

This is straight Aladura discourse. The significance of the psalms in C&S ritual comes from the evidence they offer of the defeat of evildoers by the triumphant strength of God, and the protection thus promised to his people. It is therefore the whole power battle between the forces of good and evil that is represented by the psalms. Reading these texts, in the light of the whole spectrum of efficacy embraced by the Yoruba concept of power, C&S converts could only be puzzled at the disjuncture between their powerful message and the inadequacy of orthodox church practice to protect and empower:

> The Christianity they brought to us; they told us of the miracles of Jesus Christ and the prophets of olden times, but they themselves didn't show us these things. They remained like fairy tales told to boys. They don't bring things sufficient to prove what they say; there is not enough evidence of the efficacy of prayer in what they do.
>
> (Aladura Brother Awofala)

> The CMS understand the worship of God; they don't understand the efficacy of prayer.
>
> (Elder Sister Adeoye)

> The missionaries should understand the people and the problem before they approach them.
>
> (Elder Oluwole)

The people, both in Nigeria and London, lived in a universe enlivened to a greater or lesser extent by unseen forces. The problem was the quest for both effective and protective power. It was the Aladura movement that brought the two together in a Christian idiom.

> We know that there are evil forces working against us. The natives' great-grandfathers said this, Bible proved it, and visions again prove it, so we have no other way to look.
>
> (Elder Oluwole)

The shift signaled by the replacement of *aṣe* by *agbara* in the Yoruba translation of the Bible was belied by the continuity between traditional and Aladura discourse, justified by the very text that was intended to impose a rupture with the past.

Chapter 5

"Electrical Energy": Dynamic Metaphors of Spiritual Power

Spiritual power is a dynamic concept; its significance for C&S lies in what it *does*. Those seeking the benefits of the Spirit do not necessarily worry about the way in which results appear to be produced: "We don't bother to think how or what, the purpose has to be served" (Elder Sister Sanni). "It's like an experiment. If it works, you don't wonder how" (Pastor Tomori). But this does not mean that there is no rationale for the efficacy of unseen energy. First, as we have seen, spiritual power was identified as a *distinct concept* in Aladura practice; ordinary members consulted elders specifically for prayers and techniques to augment their own spiritual strength. Second, these senior men and women felt themselves equipped to offer this advice through their own understanding of the multifaceted qualities of the Spirit. They recognized that power operates according to certain principles, for there must be predictable regularities in the way power "works" in order to put it to practical use. These meanings were constantly communicated through C&S discourse, partly in a systematic fashion in sermon or Sunday School, but chiefly through the Yoruba idiom of image and allusion in the texts of prayer and revelation. An exploration of the metaphors used to represent spiritual power, and the traditions upon which they draw, is the subject of this chapter.

Yoruba Empowerment, Old and New

#1 Brothers and Sisters; remember that the Cherubim and Seraphim can perform the wonders as of old. The power to heal, the power to make whole, the power even to wake up the dead, the power to answer what is going to happen in the future—that is our power, and is a God-given gift.

(Sunday School 23 November 1970)

The old times referred to here, as throughout C&S teaching, refers to Biblical precedents, reminding members that miracles are still possible through the power of God. But it also provides a counterclaim to the competence of indigenous medicine. A problem that much exercised elders was the confusion in popular practice, whether in Nigeria or England, of indigenous and Aladura methods of empowerment (#2, 3, 11).[1] For a generation of men and women attuned to the Yoruba life-world, even though now in Britain, the assistance of indigenous medicine was a natural recourse, and preparations were sent to students by friends and family at home (#8, 30). The thrust of much C&S teaching was the condemnation of soliciting powers other than the Holy Spirit:

#2 Herbalists, leave your juju alone;
    Come and take Holy Water

(C&S Chorus)

#3 There was this man who was coming to me for prayer; a vision had told him that he shouldn't be using medicines [i.e., *oogun*]. After some time he said that he had thrown them away, and he said I should give him some psalms to replace them.

(Prophet Korode)

Individuals who sampled the C&S in their spiritual consumer research looked for the security of replacing one spiritually charged item with another. But the continuity between indigenous ritual practice and Aladura was not based on the syncretic reproduction of ritual acts or symbols, but on the principles lying behind the concept of spiritual power, which remained substantially the same. In mid-twentieth-century Nigeria, the seeker after power was faced with a multiplicity of options from the popular practice of Islam, Christianity, and the Yoruba ritual complex. Much of the public observance of traditional religion by Christians—attendance at *orisa* festivals, attaining office in ancestor cults—was to increase personal power, whilst new potential sources of personal efficacy were eagerly explored.[2] Private practitioners, be they experienced priests (*onisegun*), diviners (*babalawo*), or entrepreneurial peddlers of power, were assured of business; the range encouraged a consumerist discrimination on the part of clientele.

This eclecticism has always been a problem for the C&S trying to pin down the "floating member" to Aladura commitment. Elders complained that "people come to K&S to compare the results with black magic and juju"; "they just go there to pass their exams, but we are not *babalawo*." This response, which combined a celebration of the Holy Spirit with condemnation of "paganism," was in line with early C&S doctrine. Orimolade's original Constitution states:

#4   The Order . . . believes in the curative effects of prayer for all afflictions, spiritual and temporal, but condemns and abhors the use of charms or fetish, witchcraft or sorcery of any kind and all heathenish sacrifices and practices.

(ESO C&S 1930: para. 10)

This Aladura baseline was reiterated in the pamphlets generated by the C&S in the early days. One of these asserts:

#5   It is certain that he who despises spiritual power and makes no endeavour to gain divine knowledge and cling to them mainly will be miserable. His hopes charttered [shattered], his labours unfruitful and his work unprofitable. For to be sure, the powers of herbs are no longer dependable . . . therefore seek these Divine Knowledges and cling unto the powers therein as they are certain.

(Sonde n.d.)

The London C&S in the 1960s and 1970s echoed the three-pronged attack on alternative sources of power evident in these texts: prohibition, scare-tactics, and claims of the superiority of the Spirit. On the first front, preaching often included denunciation, thoroughly in line with both Jehovah's and Jesus Christ's condemnation of idolatry.[3] At the 1969 dedication service of the new C&S church in Birmingham, a vision from a London prophet had this advice for the leader:

#6   To my Evangelist Olatunji: you should be lecturing the congregation several times, mostly on the difference between people worshipping God and worshipping the Devil, so that they will not disturb the prayer. This type of people, they don't deliberately want to do bad, but it is what they are born with and what they have inherited in the family. By prayer and preaching, they will change.

"The Devil" embraces any extra-Aladura ritual, not only Yoruba deities and medicine, but Masonry, Rosicrucianism, the Reformed Ogboni Society, Spiritualism—the whole array of contemporary occult practitioners. A service announcement (15 February 1970) forbade "wearing another uniform" (i.e., belonging to another society), for "if you belong to one power, then belong to it alone." The Spiritual Leader, in a sermon (4 January 1970), put it this way:

> #7 Spiritual adultery is calling another God, belonging to another body. If you want to be a K&S in the real sense—it is an unlimited church, it is an unlimited power, and it is an undiluted belief that makes you a true or pure C&S . . . Remember that adultery is not allowed in the house of God.

Exhortation is accompanied by the second strategy, that of emphasizing the dreadful dangers of "muddling up prayer and juju":

> #8 To the pregnant women, who still take the *oogun* sent to them abroad for safe delivery and are still coming for prayer to God, I don't want you to have a dead child . . .
> (Teacher Korode. Vision. Service 16 November 1969; translation)

In contrast to this, the power of prayer is safe. This point features in the third, and most prominent, line of C&S pleading—that the power of God is incomparable: God controls his own creation, obviating the need for fear, and ensuring the supremacy of the Holy Spirit (#5, 16):

> #9 As my word was over the enemies of old, so is my word today. If you want to worship the true God, we should realise that these *orisa*, they have ears but they cannot hear, eyes and they cannot see, they have mouths, but they cannot talk.
> (Brother Babatunde. Vision. Service 18 January 1970)[4]

> #10 We don't believe that witches and wizards have power over us, because we believe that our power is greater than theirs.
> (Teacher Oluwole)

> #11 Sometimes in my dream I see this man . . . He is trying to do evil, but he can't find his way through. He is using juju, but I am using my God. God makes the leaves and roots and all the things which they use, so why should I fear them?
> (Brother Oguntoye)

In all of these lines of argument, the C&S insist on *difference*, the distinctiveness of the power that is accessed through the C&S:

#12   The power which you have given to the ocean is different from other seas; the power with which you have consecrated the lagoon is different from the rivers; the power which you have given to the Cherubim and Seraphim is different from all other churches. This type of Holy Spirit, come and endow us with it.
(Prayer. Service 5 September 1970; translation)

But herein lies the tension at the heart of Aladura—one that, as we will see, was to cost it support in later years (chapter 9). The C&S must condemn all vestiges of "heathenism" in order to protect the ritual services it offers, yet its discourse of spiritual power reaffirms the existence of the entities that enliven the indigenous universe, reiterates that power may be solicited from them, and recognizes that their energy may be utilized for personal gain.

More than this, the very metaphors that represent power in ritual and revelation draw much of their efficacy from their indigenous referents. *Agbara* may have replaced *aṣe*, but language alone cannot transform epistemology. The images that implicitly convey meanings of spiritual energy in Aladura discourse carry associations not only with Christianity, but with the realm of *oogun* and the *oriṣa*. Take, for example, a common image for concentrated spiritual strength, the palm:

#13   I saw a palm leaf descended down to the soldiers of Christ Band, and they were all carrying a palm leaf. This is the spirit given to them today . . .
(Apostle Olatunji. Vision. Service 6 September 1970)

In Christian symbolism, the palm refers to Christ's triumph over death,[5] whilst palm fronds (*mariwo*) veil the entry to the shrine of the god of iron, Ogun, concealing what is sacred, excluding negative forces, and marking the dangerous energy within.[6] The conflict between Christian intent and traditional reference is built into the very images intended to communicate the basis of Aladura practice, and present continuity as a radical break with the past.

The Metaphorical Representation of Spiritual Power

The pervasive presence of spiritual power in Aladura practice has a taken-for-granted quality about it. But at the same time, the nature and efficacy of this ineffable current has to be communicated constantly, both to convince new members of its wide-ranging province and to confirm C&S adherents in their faith. The fact that this is effected more through the metaphors employed in the revelation of dream and vision, rather than through formal didactic exegesis, seems appropriate. Yoruba speech, whether ritual or secular, is full of trope and allusion, which can capture more closely the elusive nature of *agbara* than literal description. The overriding "mission of metaphor," as Fernandez puts it, is to reimagine the "abstract and inchoate" in terms of the "more concrete, ostensive and easily graspable" (1974: 123), so transferring "meaning from that which is known most intimately, very domestic understandings as it were, to that which otherwise must remain unknown and not understood" (1991: 4). To convey meanings effectively, and generate further inferences and related images to "thicken" the metaphor, these intimate assumptions must be grounded in a common culture—culture in this context referring to shared understandings embedded in language.

This multi-reference in images of power, including the tension between those drawn from a traditional or contemporary repertoire, secured their resonance with a generation educated in pre-Independence Nigeria. Some are rooted in nature, or human sensory experience. Others have cross-cultural associations, or draw on representations common to many religions. The immediacy of vision imagery, the contemporary feel of much prophetic metaphor, also derives from its reference to lived experience in Nigeria and London. But the transparency of metaphors of spiritual power to pan-Yoruba congregations containing a range of age and education is also aided by the spectrum of implied references to commonly assumed components of the indigenous culture: many echo the parables of Yoruba folk tales and *itan* (narratives), as well as ritual associations. A great number also are found in the Bible, with which everyone had familiarity through their education and private devotions. Since Biblical and Yoruba cosmologies overlap considerably, the familiarity of the images comes from their allusion to both. It was instructive that when discussing the principles of spiritual power, elders would often use the indigenous religious complex to clarify the logic of C&S practice (#25, 26, 30, 42, 43). Despite their opposition to traditional practice, their world remained vitalized by the same energy. The universality of the referents enables the metaphors to mediate between indigenous and Christian meaning, to fuse the two cosmologies around the key concept of spiritual power, and to reproduce the Aladura synthesis with every C&S vision that refers to the Holy Spirit.

The metaphorical conventions of revelation seem straightforward enough: *fruit* for blessing, *pits* for danger, *ladders* for endeavor. *Lions* display superior strength. A *pot of porridge* shares spiritual nourishment among the C&S community; *girdles* and [prayer] *gowns* provide sacred armor against evil forces, which *brooms* sweep away. There appears to be none of the opacity of Apter's ritual texts (1992), the dense allusions packed into Barber's *oriki* (praise songs; (1991), the obscure meanings of Davis's (1977) or Drewal's (1992) chants. Unlike some Yoruba ritual where fractured imagery is intrinsic to the discourse, in C&S visioning it is coherent content, not symbolic style, which is considered significant. Revelation can have a poetic quality, rich in metaphor and proverb, but purple passages are discouraged:

#14   If the Lord has sent you a message and it is only a blessing, stand up and say it, rather than telling us a lot of verbosity. We want cogent vision.
(Announcement in Service 6 September 1970)

#15   . . . if you see umbrella, you know that this is protection for the church. So say that the Lord has promised to protect us rather than giving us a lot of unnecessary words.
(Announcement in Service 5 December 1971)

This pragmatism affects the manner in which God's guidance is conveyed: omitting a description of the images actually seen in the visionary experience (as the "umbrella") in favor of their interpretation alone is a common feature in the public reporting of messages. As revelation deals in the currency of unseen forces, spiritual power is implicit in most dreams and visions, but in the telling the imagery is often left behind. In the examples of revelations quoted below, I have included only those where the visioner specifically interprets a stated image in terms of spiritual power. Even though they are in a minority, these explicit visions constantly reestablish the assumptions upon which other revelations may be understood. This is not to say that the full "fan of referents," in Turner's phrase on Ndembu symbolism (1975: 164), is consciously conveyed to every C&S member with each metaphorical message. But uncovering these referents reveals something of the coherence and complexity of the C&S understanding of the Spirit.

The Master Metaphor: Electricity

#16   The Cherubim and Seraphim is a great organisation. It is a source of electrical energy and force of power in the universe. It is a fountain of spiritual power and wisdom and understanding; it has power against the powers of darkness. It is a source of happiness, peace of mind, prosperity and long life. It is a place filled with the power of God Almighty.

(Prophet Ayorinde. Sermon 6 June 1971)

When speaking about *agbara*, there is one meta-image that members, such as Prophet Ayorinde, often use, as did their nineteenth-century forebears.[7] This is electricity, which embraces many aspects of power in action. The metaphor is modern, stressing the contemporary relevance of spiritual power, but it embraces both traditional *aṣẹ* and the *agbara* of C&S theology. Verger (1966: 36) draws the parallel between characteristics of *aṣẹ* and those of "electrical or nuclear energy." Apter (1992) has framed his discussion of *aṣẹ* with the electrical metaphor, which is also central to Gleason's portrayal of the goddess Ọya's power (1987).

A chief association with electricity is light, and—by extension—fire, the flame from wood, candles, and lamps used for illumination where there is no electric current. Light is a principal metaphor for the power of the Holy Spirit:

#17 I see the light around you, and this is the power of God, given to you today.

(Prophet Ayorinde. Vision.
Watchnight Service 5 September 1970;
translation)

#18 To the elders . . . A candle is lighted before each one of them. This is the power with which you are going to do the work.

(Lady Leader Awojobi. Vision.
Watchnight Service 5 September 1970;
translation)

In twentieth-century Western Nigeria, not only was the advent of electricity a salient marker of modernity, but light was central to the indigenous metaphor for progress: *ọlaju* or enlightenment.[8] This concept was derived from the familiar missionary metaphor of bringing the light of the gospel into pagan darkness. The allusion has Biblical legitimacy: it was light that first signaled the impact of the "Spirit of God" on primeval chaos.[9] In Christian symbolism, both light and fire have also emphasized the revelatory power of God's presence, as in "Light of the World."[10] God directed Moses from the midst of a burning bush;[11] at the critical New Testament moment of Pentecost, the Holy Spirit descended in the form of "cloven tongues like as of fire" (Acts 2:3), reflecting the diverse tongues spoken by the empowered apostles in their divine dialogue with God. This is the same spirit that is believed to inspire Aladura glossolalia and visioning, which reveals the Almighty's will. When Saul was surrounded by the "great light from Heaven," he heard and answered the voice of the Lord; he fell into a trance and saw visions.[12] So the C&S find Biblical legitimation for Aladura revelation, which replaces the Ifa oracle and priestly possession in probing the obscurity of the unseen world. The "power . . . of our Lord Jesus Christ" has provided "a more sure word of prophecy; whereunto ye do well that ye take heed, as unto a light that shineth in a dark place . . ." (2 Peter 1:16, 19). For "Thy word is a lamp unto my feet, and a light unto my path" (Ps. 119:105).

Through the agency of this powerful communication with God, the apostles accomplished many "wonders and signs" (Acts 2:43), as the C&S hope to do through revelation and prayer:

#19   Before the visioners who are raised up amongst us I see a light coming in. [If they avoid dispute and concentrate in the services] the power of God will fall upon them and they will work wonders.
(Apostle Olatunji. Vision. Service 6 September 1970)

Spiritual power stimulates revelation; dream and vision illuminate the course of effective prayer; inspired supplication regenerates spiritual power. The metaphors of light and fire speak of the circular process of communication with the divine in prayer and revelation:

#20   To our Elder Olatunji. As the prayer was in progress, I saw him in red garments. . . . and an elderly person was putting his hand on his head. As he himself is praying for people, the power is descending like fire upon him. The Lord says that the seal of prayer is being given to him today.
(Vision. Service 27 July 1969)

Visioners say that God himself is rarely seen in revelation; the "elderly person" is interpreted as angelic, or as one of the elders around the throne of God.[13] His age implies his authority to represent the Almighty in the gift of empowerment.

The color red represents heightened power; elders may wear red sashes when undertaking demanding prayer. But red also denotes danger, the destructive aspect of fire. In Yoruba daily life, fire was a real and terrifying hazard, whether tearing through the tinder of town and village dwellings, or blazing from an upturned paraffin heater, a major cause of house fires in 1970s London. Red is the ritual color of the god of iron, the *orisa* Ogun with his molten castigation; it is associated with the "hot" male power of the thunderbolts and flame-throwing violence of Sango, the god of fire and lightning. The very combination of redness and heat denotes a perilous power. Then there is the association with blood, an intrinsically powerful substance that represents life, but when associated with menstruation and childbirth is ritually destructive. "Red is dangerous. It is too hot. Yorubas run away from red; they fear it," a contemporary chief-cum-businessman told Drewal (1992: 147). "Red has 'power,' " he added, and it is able to "swallow up other colours."[14] The same reaction to red persists in the C&S. Discussing the color of the new C&S letterhead, the general secretary rejected the printer's suggestion of red on the grounds that the color was "too dangerous." A similar verdict quashed a suggestion raised in one male Band that red prayer gowns should replace white for special prayer. Although able to augment a mature power, opponents of the idea argued that to put such a weapon in the hands of the inexperienced was asking for trouble.

Redness and fire therefore denote a power that is both strong and potentially dangerous:

> #21   I saw that the whole congregation was grouped round a fire, a terrible fire . . . And I saw an elderly person standing in its midst, and we were all able to pass through this fire. And when I demanded the understanding of this fire, the Lord said that this is the Holy Spirit which is descending upon us today, and this Holy Spirit will be with every one of us, and abide with us. And this Holy Spirit will not allow any evil to pass our way.
>
> (Teacher Korode. Vision. Service 18 May 1969)

The Moral Neutrality of Power

Korode's vision presents the Holy Spirit at its most aggressive—the burning, red, and destructive "terrible" fire. And yet C&S members have the power to "pass through," to cope with it. For the destructive capacity of power is crucial in Aladura thinking. "He shall baptize you with the Holy Ghost and with fire," prophesied John the Baptist, "and (will) gather his wheat into the garner; but he will burn up the chaff with unquenchable fire" (Matt. 3:11–12). But the incineration of dross ensures a productive harvest; it is the fire of God that destroys the enemies who would wreak ruin, and so protects the faithful; it was the Lord as a pillar of fire that guided the Israelites out of their Egyptian captivity.[15] Through this punishing/protective power, the Holy Spirit, as many visions reaffirm, "will not allow any evil to pass our way." In Yoruba myth, it is the fiery gods Sango and Ogun who represent the righteous wrath of God. Thus Sango brings blessings through his persecution of transgressors;[16] Ogun's devastation clears the path for regeneration.[17] To his devotees, this violent orisa is the "giver of peace and preserver of lives" (Akinrinsola 1965: 85). Thus fire imagery in revelation indicates the dual significance of the exercise of power; the fatal fire of divine wrath evokes the possibility of life.

The ambiguity of the image mediates contradictions that are intrinsic to the nature of unseen power; it carries both a positive and negative charge, according to how it is employed. Since the predominant moral message of both indigenous myth and the Bible is that good can conquer evil, the beneficial potential of power is paramount, but spiritual power in itself cannot be characterized as good or bad.[18] It carries the capacity for both:

#22   Power is like electricity; it gives you light, it also instantly kills you.
                                                                    (Pastor Tomori)

It is this amoral nature of power that accounts for the capacity of all indigenous spiritual entities to help or harm. The character of an orisa is a twosided coin: the fiercest deities can bring blessings to their devotees. As we have seen (chapter 4) this even applies to witchcraft: "Where there are bad things, there you get good things too," said a Yoruba, dancing in honor of the witches (Thompson 1970: 75). Aladura often attributed space travel and scientific discovery to witchcraft and "means" (i.e., juju) as well as directly to the power of God. The missionaries equated the orisa Esu with the Devil, but in the indigenous cosmology Esu was more a spiteful, unpredictable trickster than an embodiment of evil.

The C&S have adopted the orthodox Christian view of Esu, but the moral neutrality of power persists as one of the most resilient indigenous legacies to the dynamics of C&S ritual. So prayer, which is essentially the mobilization of power, is described as a weapon—a gun, a knife, or a sword, which can be used to attack as well as defend, "an instrument that you can use for any purpose you like":

#23   There is prayer that can cure you and at the same time here is prayer that can kill you. There is prayer that can help you; there is prayer that can harm you. There is prayer that can assist you, and there is prayer that can put impediment in your way.

(Pastor Tomori)

The motive may be retaliation rather than malice:

#24 It is not hard to use prayer for destructive purpose. If you want an enemy to have motor accident—two minutes—you can finish him now—yes, killed—just like that!

(Pastor Tomori)

But victory prayer (*adura iṣẹgun*) is discouraged:

#25 Don't pray for conquering somebody who is running after you. You have your God, and one with God is a majority. So don't go about praying *adura iṣẹgun* or anything like that.

(Announcement in Service by the Spiritual Leader 22 February 1970)

In spite of such ethical restrictions, the Aladura equivalent of cursing is still an option, for the moral ambiguity of *agbara* is fundamental to the economy of spiritual power and the significance of ritual symbol in the C&S.

Power as a Comparative Concept

#26 I was told of a dream experienced by a prophet while he was still "an *Egungun* person"[19] which assisted in his conversion to Christianity. He saw a tall pillar, composed of different coloured sections "like a rainbow standing up," the first layer representing "the power of this juju," the next, "the power of another juju," and so on. "It continued on, grade by grade, right up to Heaven, further than you could see." Then he heard a voice telling him that "this is the power of prayer, for there is no end to this power, and it is above all others."

Power is a hierarchical concept; differing entities are endowed with the amount and type of power appropriate to their status. The spiritual strength of C&S prayer is claimed to be incomparable because it provides access to the ultimate power in the universe: *Oluwa Ọlọrun Olodumare* (Lord God Almighty); *Alagbara julọ ninu awọn ọrun* ("The most powerful one in the Heavens"). As he is the source of all spiritual energy, whether of *orisa* or angels, he is *Ọlọrun agbara ti i gba gbogbo agbara lowo mu*—"God the power that handles (i.e., dominates) all other powers". This is why he is *Alaba asa si* ("The Lord with power of protection"). These names are among those used to address God in C&S prayer, but they are inspired by the praise traditionally offered to Olorun. For by retaining the indigenous term for God, the translators of the Bible effectively promoted religious continuity; God was reimagined, but not replaced.[20] Just as the Archangel Michael routed the Devil for "salvation and strength" (*igbala ati agbara*, Rev. 12:7–10), so in both indigenous and Christian traditions, God can ultimately control all other entities, including Eṣu, because it is from him that their power derives (#10, 11).

Although there is a rough ranking of spiritual and living beings and material objects, position between and within each category can fluctuate; the *emi* can be supplemented by manipulating a lesser strength and tapping a greater vitality. In this battle of strength, the aim is to overpower others:[21]

> #27 A lot of spiritual things are a matter of one spirit being stronger than another. So with *oogun*, the spirits you call up may be stronger than the spirit of the person you want to influence, and so you will dominate him. A lesser power must always submit to a higher one, unless it is being used by a greater spirit.
>
> (Prophet Ayorinde)

In this cosmology, the template for Aladura ontology, there is a recognized hierarchy of *ase*—albeit an unstable one—among the dominant *orisa*, with lesser deities, witches, ancestors, spirits, and medicine following on behind. The whole cosmological structure is articulated by power, begged, borrowed, or commandeered from greater or lesser sources. The *orisa* constitute the main channels of God's power, yet their myths show them tapping the powers of medicine, as do their priests. Ancestor cults enlist the help of "our mothers," and *onisegun* placate them, for they can destroy the efficacy of medicine. *Onisegun* also petition the power of the deities, but conclude their preparations by praying: "*ki Olorun ṣe eyi to ku*," "God will do the rest," for, ultimately, the power of each medicine is given by God.

This hierarchical conception of power is critical to Aladura practice. As a teacher said, "If you have a greater power, the other will be bowing down for you." Adequate power is not just a bonus ("enemies will fear your energy"), it is essential. A brother told me: "People will test your power; if you do not have enough power, they will kill you." Not only this; anyone who attempts to tap the power of another source needs to be up to the task; anyone handling electricity without adequate protection and knowledge can suffer shock, hence the reservations about brothers' red prayer gowns. For unless strength and experience is sufficient to control the spirit summoned, the effects of the unleashed power can literally, it is claimed, blow up in the invoker's face. This was the explanation given by an elder when an unfortunate new member placed a red candle on her television to boost her prayer, and the set exploded!

The danger of recoil is greatest when power is activated to harm another: the witch with the gourd was cautioned so that the blood would not run over her own head (chapter 4, p. 69). What Maclean calls "the boomerang principle" (1971: 95) abounds in Yoruba myth and anecdote, where humans and gods alike are hoist with their own petard:

> #28   All these things work on a spiritual plane. If the spirit called up is
>   stronger than your own spirit, then it can overcome you, and you
>   can be running mad because you think that you will use them,
>   but they will be using you.
>
>   (Prophet Fakeye)

Just as there are said to be medicines and incantation that can cause power to backfire,[22] so prayers ask for the same effect:

> #29   For all those who are praying against the success of the anniver-
>   sary of this great Band, O Lord, may it go back on them.
>   (Prayer. President of Fogo Band.
>   Anniversary Service 16 October 1970)

But power may rebound spontaneously. This occurs when the target is too powerful, when power runs amok, or when correct procedure to control unleashed power is neglected. It also happens, members explained, when the assault is unjustified, when a curse (*epe*) is unprovoked. Although the previous exercise of spiritual energy was ideally tempered by justice and moral values, Peel argues that it was Aladura who consistently attempted to subject *agbara* to morality (1968: 138). It is certainly the moral aspect of rebounding power that is stressed in C&S: perhaps the victim was in a state of impurity, or their *emi* "ungodly." A woman visiting a prophet for healing complained to him that not only was his Holy Water ineffective, but that it had given her a headache. He privately concluded that it was her own witchcraft that had caused the water to "turn against her."

The message that malice brings its own reward was the subject of this sermon:

#30   Two Nigerian couples were living in a house in North
       London . . . sharing a gas cooker. It happened that the wife of
       these two men were fighting because of "I slot [i.e., put money]
       in the gas meter; you did not slot in the gas meter." The other one
       was so unchristianly, was so ungodly, she said, "You are having
       it!" [i.e., going to get it]. She told her husband, and he wrote
       home [to Nigeria]. And one man replied him, and said, "I will
       send you one cudgel.[23] When the cudgel comes, as soon as you
       have a quarrel with the man, give him two or three blows, and
       what will be the end of this man, he will be sorry!"
            The other man was working in the Post Office . . . Any letter
       which came [for his adversary] he would open it, and then tear [up]
       the letter. When the man found that this cudgel was coming for a
       purpose, he took a sick leave from the Post Office. When the cudgel
       came in a parcel with some charms, it also said how to use the
       cudgel, so he started to recite these charms. When he knew that he
       had mastered them, he instructed his wife to go and fight the other
       man's wife again. Then he gave him some lashes, and immediately,
       the man [wielding the stick] was mad. And the woman [i.e., his
       wife] started to complain: "Ooh, we have done ourselves! Oh,
       whatever have we done?" Then the other man said: "You were the
       one who sent for the cudgel; you never knew that this would be the
       outcome!" So the man was sent to Surrey [mental hospital], and
       because he could not be cured, he was finally sent [back] to Nigeria.
                                          (Sermon 7 December 1969)[24]

The congregation found this cautionary tale highly amusing, but it contained a salutary message: recruiting the power of other ẹmí without the necessary personal strength to handle it is either dangerous or useless.

Personal Spiritual Development and the Control of Power

The development of individual spiritual energy is also of prime importance in reaping the benefits of the power of God:

#31   To the Choristers: I was made to see a certain flower descending
       unto some of them. Some of them are planting this flower in
       order to make good progress and also to go further. But some
       people are only putting it in their clothes. I demanded the
       understanding, and the Lord said that these are the blessings and

the glory which are given to this Band. Those that are wise enough plant this thing, but those who are not wise enough are just handling it. Those that are planting theirs will go further and further. For this thing is a symbol of blessings and spiritual power, and those that are talented to acquire this should pray, fast, and acquire more spirit.

(Aladura Brother Oterunbi. Vision. Service 7 December 1969)[25]

Although a flower is not a familiar traditional symbol in Africa,[26] fruits and plants—like the palm—have resonances of spiritual power. In C&S revelation, fruit is a common metaphor for blessing, a concept that refers both to God-given successes, and directly to the spiritual power through which these are bestowed. But plants, like personal power, need tending; "When you plant a tree, do you cover it? When you plant a flower, do you hide it away?" asks a vision urging an elder to develop his power of "vision and dream." "Be careful, you must cultivate them, or they will die" (Service 20 April 1969). He is counseled to protect the plants with a fence lest they be destroyed, that is, by wild animals (evil spirits) or power-predatory enemies. But loss of power may be as much the result of personal negligence as spiritual attack. So a visioner saw:

#32   ". . . certain trees shedding all their leaves." The trees had been planted "for them to make good fruits, to feed them to my people, and for them to give shelter to my people during the time of hot weather," but they were "not fulfilling their purpose." The Lord therefore warned those who are coming to church merely as a "spectator or watchdog," that "their powers are being diminished."

(Aladura Brother Oterunbi. Vision. Service 7 December 1969)

Spiritual weakness undermines ritual efficacy. C&S practice aims both to generate spiritual energy and to harness it to a pragmatic end: to make prayer "work." This requires personal strength to handle power. The "elderly person" appearing in visions 19 and 20 speaks of this necessity. The white garment in which these elders appear represents their capacity to contain surrounding forces, as can white-robed Biblical angels, and those " 'cool' white *orisa* of female composure and control" (Apter 1992: 64). The theme of containment and the necessity for control, together with the dangers of overflow and excess, are central to indigenous conceptions. Potency contained is animating, but unfettered it destroys. So with spiritual power, whether of medicines, deities, or the Aladura God. Power that is overwhelming, that cannot be handled, is dangerous, hence the importance of prayer and fasting, building up this capacity for mastery. A vision directed at a newly consecrated prophet in the 1970 ordination instructed him to pray that his own strength would be sufficient to make use of the power he had just received. For spiritual strength is a balance between the gift of the Holy Spirit and personal effort:

#33 So your own spiritual power can increase and decrease, and the part of the Holy Spirit in you can increase and decrease. It's like a car battery, which has acid and distilled water. What happens if there is no more acid? The car doesn't work any more. What happens if the distilled water is all dried up? The same thing. Then let's say that the acid is supposed to take three-quarters of the battery, and the water one quarter. But if the acid is only a half, and the water is not proportionate with how it is supposed to be, the battery will be very weak. Then it needs charging, and you put it on charge. That's how the Holy Spirit is, and that charging is your prayer and fasting.

(Prophet Korode)

Metaphors of modern transport for spiritual energy vividly convey the significance of development and control:

#34 To Elder Ojo. The Lord says that he will be blessed in spirit and naturally . . . There is an aeroplane before him in the New Year. He should be praying to be able to use it.

(Aladura Brother Oduntan. Vision.
Watchnight Service 7 November 1970; translation)

#35 And before our Brother Oguntulu, the Lord is asking of a helicopter, a train, and trekking, which is going to be faster? Since the last eight months the helicopter which God has given him to ride on has been changed to a train. That is slow progress, and the helicopter was even expected to change into an aeroplane. In spite of all this trouble, God is still wanting to help him. But he must be very prayerful that the train is not changed to trekking.

(Elder Oterunbi. Vision.
Watchnight Service 9 January 1971; translation)[27]

Otherwise, as Oguntulu's vision warns, power diminishes, and the driver of a potentially powerful vehicle may find himself plodding along on foot. But these visions do not only address the necessity of control, they stress its potential: the possibilities inherent in handling power, the partnership between vehicle and owner, the exhilaration of successful mastery.[28] This is a different message from the threat of dwindling efficacy (#31, 32, 33), or insistence on the process of prayer, fasting, and patience. It emphasizes the reward as much as the obligation of spiritual development:

> #36 At the beginning of the service, I saw people riding motor cycle, very powerful riders . . . They are so fast, it is just as if they are flying in an aeroplane, and they were sent today to bring the power of God into the fold.
>
> (Aladura Brother Babatunde. Vision. Watchnight Service 9 January 1971)

In Western Nigeria of the 1960s, mechanized modes of transport were major symbols of modernity, whether of the nation boasting contemporary systems of communication, or of individuals proving their economic success. Such images stress the speed and force of spiritual energy, its capacity to defy gravity that grounds most objects to the earth, to supersede the laws of nature. It is volatile, dynamic, suited to the change and movement of members' lives. Metaphors from modern technology, such as the image of electricity itself, address the possibility of human control of the environment through means of the Spirit, and the competence of spiritual power in a changing world.

Secrecy and Disclosure

Metaphors of modern transportation are devoid of ritual reference and depend on contemporary experience. For Aladura of the worker-student generation, this experience was one of dramatic social change, whether of the colonial encounter, urban migration, or voluntary exile overseas. Common to all of these was confrontation with the *oyinbo*, the European, and with the economic and political dominance of the colonizers. Comparisons, in the early days, were invidious. The unseen power of indigenous medicine was said to transport through space, or enable communication across distance; yet here were visible machines that produced the same result. It may well be that conclusions were drawn about the spiritual secrets behind their power.[29]

Secrecy is a subject that much concerns the Yoruba; what is hidden, safeguarded from the common gaze and from the potential pollution that would dilute its strength is considered intrinsically more powerful than that which is open and revealed. Part of the potency of spiritually charged symbols, be they from the indigenous or Christian repertoire, is the hidden, enclosed quality of the power they contain. The ritual specialists in divination and esoteric knowledge are known

as *babalawo*, "father of mysteries" or "father of the secret things."
Hence the potential attraction to Aladura of occult literature such as
The Original Key to the Sixth and Seventh Books of Moses or *The Greater
Key of Solomon*, books that are believed to have been taken out of the
Bible—some say by the missionaries—because of the contact with
supernatural powers they provide (see chapter 8, pp. 197–8)

The image of the key in these titles refers to closure, concealment,
to personal, privileged access to something desired. The key is not a
familiar metaphor in indigenous ritual texts or iconography, for keys
were not in common use in precolonial Western Nigeria. But once
introduced, padlocks became incorporated into the practice of juju,
especially in matters of sexual fidelity. In secular daily life it always
struck me how often church members in London kept the doors of
their rooms locked even in their own house, especially if there were
other tenants, protecting themselves against intrusion upon their space
and belongings, or into details of their closely guarded personal lives.
Even if the key in visions conjures up something more magnificent
than a small brass Yale, the image still carries meanings of this exclusion
and secrecy.

But the key is also a metaphor for opening and revelation.[30] The
occult texts, popularly known as *Six and Seven*, are condemned by the
C&S, for the church's emphasis is on the democracy of the Spirit.
There is tension here, between the official line promoting prayer and
spiritual development and the penchant toward the acquisition of esoteric formulae. The image mediates between clandestine occultism,
and the popular availability of *agbara emi*:

> #37 To the leader of the service, and to the executive committee. The
> door which I have opened for you, I have given you the key. Why
> are you afraid of entering into the house I have provided for
> you? . . . The door is the blessing with which I have endowed the
> church, and the key is the spiritual power which I have sent down
> for the church . . .
>
> (Interpretation by an elder of a brother speaking
> in tongues. Service 21 December 1969)

The key to God's house also speaks of the passage from paganism to
Christianity, and from orthodoxy to Aladura. Just as the missionaries'
gifts of keys and padlocks signified widening horizons as well as enclo-
sure, so the revelatory key represents the transition from a state of
spiritual immaturity into a new dispensation of power. This access to
the divine is seen to lend the church its authoritative efficacy; Christ's

promise to Peter is reiterated in sermon and prayer as relevant to the C&S:

> And I will give unto thee the keys of the kingdom of heaven: and whatsoever thou shalt bind on earth shall be bound in heaven: and whatsoever thou shalt loose on earth shall be loosed in heaven.
> (Matt. 16:19. Cf. Isa. 22:22 and Rev. 3:2)

The Authority of Power

The message of the supremacy of God's power over all others is conveyed by metaphors of prestige. As a metonym for kingship, the significance of the crown in C&S revelation is status and authority:

> #38   I saw three crowns before us. We should be very prayerful, because the Lord says he is going to elevate three people in the fold between now and 1971, to a very high standard post in life.
> (Prophet Aiyegbusi. Vision.
> Watchnight Service 7 November 1970; translation)

The Biblical precedent for this metaphor is vivid. Take, for example, the Book of Revelation, where the contest between God and Satan is played out in the familiar visionary idiom. The crown is a central image, which stands for the power of all adversaries: the mother of God, in her starry crown, is menaced by Satan, "that great red dragon, having seven heads . . . and seven crowns upon his head" (12:3).[31] The crowned beast of blasphemy, empowered by the Devil, confronts the "saints" (chap. 13), but the "King of Kings," the "white horse" whose "eyes were as a flame of fire" and on whose head "were many crowns," claims the ultimate victory (19:11–12, 16). The faithless who have succumbed to Satan are accordingly attacked in the "bottomless pit" by terrifying creatures who were "given power as the scorpions of the earth have power" (9:2–3), and "on their heads were as it were crowns like gold" (9:7). The crown is power, and on the head of legitimate authority—even if used to torment—represents the triumph of God over the deadly forces of evil. "Be thou faithful unto death," quote the C&S from Revelations (2:10), "I will give thee a crown of life."

But it is from the traditional political system that the image has acquired its richest resonances. Although the scope of indigenous office is now much restricted, crowns are still made and worn, and both the nature of kingship and the royal regalia were familiar to Yoruba in London. In the precolonial era, and still to a certain extent today, the rulers of Yoruba kingdoms were religious as well temporal leaders, patrons of local *orisa* cults and ritual societies.

Nowhere is this theme of the interpenetration of political and spiritual power made more visible than in the beaded crowns (*ade*) of Yoruba *oba* (kings), their elaborate iconography constituting dramatic statements concerning the complexities of royal power.[32] As the dominant image of kingship, these magnificent, colorful structures, with their bricolage of symbols brimming with power, proclaim the king's inner efficacy, and the potency of his office.

Power and Transformation

Another such image is that of the staff or rod (*opa*), an item of royal regalia usually carried in front of the *oba* or his messengers, representing his delegated authority. In indigenous ritual, *babalawo* and *onisegun* wield rods to indicate their God-given powers. In C&S ritual settings, evangelists, pastors, prophets, and apostles hold short silver staffs or crooks, made of wood or metal, to represent their power.[33]

> #39 Evangelist Olatunji . . . There are a lot of blessings before you, because I saw three rods, which can do different things, which is the Spirit given to you in the service of today.
>
> (Teacher Aiyegbusi. Vision. Revival Service 18 May 1969)

Salient among the "different things" effected by the power of these rods is the capacity to deal with the ungodly powers broken by the rod of the Almighty[34] or defeated by Moses and Aaron's staff.[35]

> #40 To the leader of the service. I saw you given a certain rod in a bush-like place . . . This is the power of the Holy Spirit given to you today.
>
> (Sister Akinwale. Vision. Revival Service 18 May 1969)

The "bush" carries associations with dangerous, uncontrolled indigenous spirits, which can be countered by the power of the Spirit. So the familiar words of Psalm 23 assume an aggressive significance in Aladura discourse: "thy rod and thy staff they comfort me," not merely by providing the guidance and support of the Good Shepherd's crook, but by defying those ever-present enemies.[36] A staff both supports the feeble and smites down the foe; it attests to the moral ambiguity of power. This mediation between conflicting attributes, the transmutation of negative into positive, the elision of benefit and harm is present in all spiritual agency. In the indigenous cosmology it is Eṣu, the trickster deity who specializes in the moral flexibility inherent in power, who carries a stick or club.

Eṣu also specializes in transformations. The power to assume another form, or effect such a metamorphosis, is a critical part of Yoruba medicine, and lies at the heart of the witchcraft complex. Yoruba mythology is full of stories of transmutations.[37] This transformative power of the rod is a salient theme in C&S discourse. Not only did that of Moses and Aaron work many "signs" of turning one thing into another, but the rod itself changed into a serpent when cast onto the ground.[38] When C&S elders dip their staff into water, it is transformed into a power-charged substance. By the transference of a portion of God's power, the material is invested with spirituality.

Many C&S members accepted the possibility of physical metamorphosis, claiming to have witnessed men turning into animals, or recounting anecdotes that circulated in their childhood. Later they came to see this display of power as unchristian juju, which smacked of witchcraft. But on a spiritual level, transformation is of great significance to the C&S. As Christians, Jesus's death on the cross is recast as everlasting life; as Aladura, the limitations of human consciousness are superseded by prophetic spiritual awareness, and the mortal visioner is transformed into a mouthpiece of God. The possession priests of the *orịṣa*, Ogun and Sango, have ritual staffs; in the Old Testament, Jonathan's rod enabled him to see visions;[39] the rods carried by C&S apostles and prophets signify their ability to surrender their own personae to the Spirit, and to replace earthly sight with "the spiritual eye." Even if an apostle does not become possessed, his silver rod refers to the transformative power of effective prayer, which can convert danger into blessings, and enable the change of fortune that every member desires.

Reification of the Unseen

But the rod does not only speak of power or act as a convector, it embodies it. So also the crown: hidden inside royal headdresses are powerful medicines; incantations accompany the preparation of crown and staff for use. These objects both indicate power and contain it. Through the appropriate ritual, they are transformed into inherently powerful objects: spiritual energy given physical form. The crown can thus render the king inviolable, but pointed in anger at an opponent it becomes a lethal weapon.

This transformation addresses an apparent paradox in the nature of spiritual power itself. Although intangible, power is not just a vague, freefloating essence. It is also reified, a "thing" to be transferred from one person to another. Take this passage from a contemporary short story concerning an itinerant Prophet, Lucky, whose bizarre behavior provoked a church congregation to eject him from a service:

> Lucky screamed at them! He bubbled with power, it was frothing out of his mouth. He cursed them!
>
> (Osofisan 1991: 24)

It is not only his words that are intrinsically powerful; Lucky himself is overflowing with the substance of power. Apter describes the sacred calabash of the goddess Yemoja, "heavy" with her *aṣe* (1992: 105). Moses transmitted his spirit to Joshua;[40] Elijah empowered Elisha by bequeathing him his *emi*, his spiritual power:[41]

> #41   The Lord says that he will talk to you in a proverbial way. There once was a king who was planning to go on a long journey, but he fell ill, and called the doctor. When he came, the doctor said that he would cure the king, but only if the king gave him his crown. Secondly, the Lord said that there was once a young girl going to her husband. But she too fell sick. The doctor told her he would cure her in exchange for her golden ring. Thirdly, an Apostle of God became sick, and the doctor said, "I will cure you if you give me your spiritual power."
>
> If the three of them have faith in God, the healing power can come from God. The king, bride and apostle are the members of the C&S. If they have no faith and don't believe, they will sell their talents to the enemy. But God has promised not to take away their power, the crown, and the ring, if they turn to him.
>
> (Summary of the English translation of a vision read out to the congregation. Service 20 April 1969)

In this vision, the doctor (i.e., an *oniṣegun*, skilled in indigenous medicine, for C&S do not oppose biomedicine) is demanding that C&S members hand over their spiritual power in some Faustian pact. This theme is reiterated in the Biblical text often cited by C&S, in which Simon, "who was a witchcraft, and used evil power" (Sermon 4 January 1970) solicited the evidently superior spirit of Apostle Peter in Acts 8:9–24. Peter rejected the idea that "the gift of God" may be purchased with money (v. 20), "for thy heart is not right in the sight of

God" (v. 21), so counterpoising the material transaction of juju with the ethical qualifications necessary for the Holy Spirit. But this is a matter of the morality now associated by Aladura with spiritual power, rather than a shift in the nature of power itself. Like electricity, power is a process, a current, not an entity. Yet Morton-Williams (1964: 250) has described medicines, made with the help of the *orişa*, as "storage cells" of the deities' power. To pursue the electrical meta-image, Yoruba ritual symbols, whether traditional or Aladura, are likened by elders to "batteries," receptacles of power, which generate access to refractions of God's spiritual energy. C&S discourse reveals that invisible power still may be conceived of as something almost tangible, which may be illicitly traded as well as legitimately acquired. It can be both contained and stored in sacralized ritual objects, words, and gestures. In turn, these vessels of power contain the capacity for generating further power, or releasing the vitality latent in other creatures and things.

There remains the question of *how* both sentient and nonsentient beings become infused with power: does the whole universe pulsate with innate energy, or are the mechanics of empowerment more precise?

#42   Some people believe that in any existing thing there is a force behind it. The Bible explains that God has created things, and if the things created have force, then other things will have force. Human beings were created last, therefore they have superior force. Water is having its spirit, so is stone, because it is created. Leaves and plants too, they were created; they have their own power, so can be used for so many things.

(Prophet Korode)

This understanding of Genesis 1 accords with indigenous metaphysics, whereby the elements, animals, plants, and human beings are empowered by creation. These are categories "wherein there is life" (*ti işe alaye*) (Gen. 1:30) and therefore power. As to other nonsentient material, C&S elders tend to echo tradition in emphasizing spiritual agency rather than innate energy:[42] things become empowered through deities and spirits either on the volition of these entities, or because they are ritually summoned to do so: "An object has power because a god has come in after someone has dedicated it," said Pastor Tomori:

#43   Look at fire burning! You know that normally it will burn you. Why? Because there is a spirit in it forcing it out, a spirit controlling it. But if you are powerful and you know how, you can call the name of the spirit to quench it without water. So with a gun, you can ask the spirit to stand still and it will stop the bullets, or they will come out and not hurt you, like turning to water. You do this by calling on Ogun, the god of iron, you call on this spirit. This does not mean that gun has spirit, but that there is a spirit guiding it. It has got spirit from where the force is coming from, and you have to know the way to call it and get it to act.

To render items ritually significant, God must thus be requested to energize objects that are "dead," or whose level of power is low (#33). This is achieved through "dedication," prayer for "sanctification," or "consecration," all referring to the process of empowering with *agbara emi mimo*:

> #44 The Spirit of God will only descend in a clean place, so you have to consecrate a new prayer gown, for example, against the evil spirits to make it clean. Then God will move near, and the power will be greater than ordinary if you do this. Consecration makes things clean, and clean means power. A consecrated prayer gown is powerful because power is now inside it.
>
> (Prophet Korode 10 May 1970)

The Yoruba term *mimo* means both clean and holy, as in the Biblical concept of ritual purity. Once consecrated, the white garment becomes a vehicle for the Holy Spirit, transformed into armor against evil forces, while attracting the angels "who will help you all along," aiding the efficacy of prayer. Elders point out that "it is not the dress itself which is powerful, but the spirits which are called up." The average member will not distinguish between inherent vitality, or the greater access to the Holy Spirit, which the white uniform affords; what matters is that a blessed gown is "very useful." But in C&S orthodoxy, consecration is a critical process; it animates ritual speech with performative power and imbues people, places, and ritual symbols with spiritual energy.

The Power of Precedent

The metaphors representing spiritual power draw on both Yoruba and Biblical traditions, fusing concepts of *ase* and *agbara* to forge an Aladura understanding—Christian, but invested with culturally familiar meaning. References to the past also exemplify the particular importance that Yoruba place on precedent. Saint Peter's power, for example, is mentioned in visions, not just for rhetorical effect, but because it is seen as a model for Aladura empowerment. The "wonders as of old" in the Sunday School text (#1) are recalled not only to provide specific Christian legitimation for C&S practice, but as historic precedents for the power of contemporary prayer. Just as electronic formulae, once proven, will produce standard results, given similar conditions, so the efficacy of prayer is ensured by previous success. This concern with case history, as elders often call it, runs through indigenous culture; in their *itan*, their oral histories, Yoruba constantly bring the past into the present in order to create the future.

The significance of precedent is legitimated by revelation:

#45 God is providing more for the barren women . . . As I have gone to Sarah, and fulfilled my promise to Hannah, so I am in your presence today.
(Aladura Brother Babatunde.
Vision. Service, 18 January 1970)

#46 As my word (prevailed) over the enemies of old, so is my word today . . .
(Aladura Brother Adeoye.
Vision. Service, 2 November 1969)

#47 When I called Peter, I told him to leave everything he's got and follow me, and I will give him everything back . . .[So] don't worry about exam more than your God.
(Teacher Korode. Vision.
Revival Service, 1 February 1970)

These visions remind the congregation of God's former blessings; prayers remind both members and their God of previous events. Thus prayers begin "Lord God of Abraham, Lord God of Jacob, Lord God of Isaac . . . ," the names providing metonymies for particular displays of power. Or they refer to the miraculous days of the early C&S: "O God of St Moses Orimolade Tunolashe . . ." References to Old Testament characters, as in the Watchnight consecration prayer (chapter 4, pp. 63–4) are scattered through supplication:

#48 . . . the type of spirit which you gave to Moses [Elijah/Joshua/ Samuel/Isaiah . . .] come and descend it in our midst . . .

#49 O Heavenly Father, Daniel knelt down before thee and he said, "What will the king ask me?" You did not allow him to fall . . . Here we are, the Cherubim and Seraphim. We are all students, we are kneeling down today that when we shall be sleeping tonight, teach us what the examiners will ask us; as Daniel stood

before the king and he was able to foretell, and the thing came to
pass . . .

(Service 1 June 1969)

The Gospels also provide precedents: in a prayer during Lent, God
was reminded that the power given to Jesus after his 40-day fast was
so great that even the crucifixion "had no effect on you." This principle
of precedent lies behind the use of psalms and portions of the Bible in
prayer for particular circumstances: "You know that somebody has
spoken it and has got the answer," said the pastor. He quoted the
story of David and Saul (1 Sam. 24) as relevant to the betrayal of
friendship, and commented:

#50   You know that prayer is just like a law, like a certain legal history,
       then you've got to cite the history, and the government will give
       you the day. You have to remind the angels that this is the same
       thing as this man is doing to you, and that the same judgement
       must be served. Then you will see the effect—just like that!

In indigenous ritual too, the gods are reminded of past instances of
their powers. For the preparation of *oogun*, incantations (*ofo*) cite the
proven efficacy of the ingredients in order to activate their vitality. Ifa
divination depends on the quotation of oral literature featuring a
precedent for the client's predicament (see chapter 8). The parallels
between *ofo* and a prayer charged with historical precedent and praise
names are not lost on C&S elders. They explain that by prevailing
upon the controlling spirit to act as in the past, ritual efficacy is
increased in both instances, for the principles upon which the electrical
force of spiritual power remain substantially the same.

Despite the apparent simplicity of C&S images for this vitality, they
represent a complex concept, full of ambiguity and contradiction.
Binary expression is intrinsic to the Yoruba language, and so with the
tropes it employs; metaphor can both convey disparate meanings
simultaneously and mediate between apparently irreconcilable elements. Spiritual power is associated with the sacred and the secular;
divine supremacy and individual agency; secrecy and revelation;
knowledge and faith; male potency and female spiritual energies; historical depth with contemporary immediacy. It signifies good and
bad, life and death, danger and security. It is at once elusive and definable, intangible and capable of being transmitted and contained.
Power is God-given, "natural," yet must be developed and controlled
for human use. For power is never static; it is concerned with

movement, transformation, sliding between interpretations and meanings. Revelation and prophecy convey multilayered messages on these meanings beyond that of their manifest content. They communicate aspects of this central Aladura concern on the conscious or unconscious level of understanding to which metaphor can speak.

Chapter 6

Experiencing Power: Possession by the Holy Spirit

Although the ritual texts of prayers and visions communicate the concept of spiritual power, it is not words alone that convince the members of C&S of the reality of the Spirit and of its multiple meanings. It is when the power of the Holy Spirit possesses the bodies of the spiritualists that these understandings are underscored by personal experience. Although much has been written on spirit possession in small-scale societies, this has rarely been related to the study of the Spirit in AICs or Pentecostal churches. The case of Aladura, emerging from a Yoruba ritual tradition into Christian monotheism, invites comparisons. It also suggests connections between developments in the conception of the possessing power, and of individuals' experience of their changing society.

The Watchnight Service

Six o'clock on a Saturday evening. The pews in the Earlham Grove Church Hall have been pushed back, leaving an empty expanse of lino for the free movement of the Spirit. All services, all prayers, provide an opportunity for a spiritualist (*elemi*) to experience possession by the Holy Spirit—but tonight is a Watchnight, a service set apart from the mundane activities of daylight hours and devoted to the descent of spiritual power and revelation from God.[1]

Singing, we process somewhat raggedly into the hall, the sisters arranging themselves in lines on the left-hand side, the brothers on the right. So far there are only 60 members present, but more will arrive during the next hour until there are some 50 women and 55 men in the congregation. The senior elders—three apostles, three prophets, and three evangelists—bring up the rear, to stand in front of the altar platform. Lastly come two elder brothers, the first carrying a large wooden cross. He is followed by the leader of the service, Elder Bayo Oterunbi, a stocky man in his mid-thirties, now completing the final stage of his accountancy course. Watchnight service leaders are not invariably spiritualists themselves, any more than are senior elders— two of the apostles and one of the evangelists do not become possessed. But Oterunbi's father is a C&S apostle in Nigeria, so he himself has been a member "right from home," where he first experienced the Spirit; now he is a prominent visioner in the church. He faces the altar and recites Psalm 24 three times. We all join in the third time round:

> Lift up your heads, O ye gates; and be ye lift up, ye everlasting doors;
> and the King of Glory shall come in. Who is this King of Glory? The
> Lord strong and mighty, the Lord mighty in battle.
>
> (Vv. 7–8)

The man responsible for conducting the service, in which the all-powerful "King of Glory" will enter into the receptive bodies of spiritualists, needs strengthening himself: Oterunbi kneels as three of the senior elders pray for him. He then stands and turns to face the congregation to pray with Psalm 101, verse 7, for all the "visioners and dreamers." The former are few, perhaps a tenth of the congregation, even though two or three times that number will show signs of possession. But any active member can credibly claim the power of dreams (*ala*), and all kneel as they are warned against concocting false revelations:

> He that worketh deceit shall not dwell within my house: he that telleth
> lies shall not tarry in my sight.

Oterunbi takes the cross, mounts the platform, and lights the seven candles on the altar, while an elder calls out the lines of the next hymn. "Clap your hands," shouts the Spiritual Leader, Apostle Abidoye. "Clap! This is very essential for falling in Spirit!" Lady Leader Tomori turns to face the women, clapping enthusiastically, with particular encouragement to the "spiritual" ladies. The clapping intensifies; bare feet slap the floor as the white-robed members stamp and jump. There is no organ or drumming in Watchnights, but Evangelist Somefun, a young and popular visioner, beats out the rhythm with the clapper of the handbell. Elder Olusanya, a small, sedate man in his fifties, who has settled in England as a practicing GP, wafts incense around the hall and over the congregation to expel evil spirits and open the way for the protective power of the Almighty. Already power is making its appearance. An elder sister, oblivious to her surroundings, is clapping and dancing to her own rhythms;[2] several heads are shaking, bodies trembling as spiritualists move into another consciousness. Oterunbi continues with the readings: Psalm 51 for "Forgiveness of Sins," which could obstruct the descent of the Spirit, while Isaiah 6 summons a purifying power in preparation for prophecy:

> Then flew one of the seraphims unto me, having a live coal in his hand, which he had taken with the tongs from off the altar: And he laid it upon my mouth, and said, Lo, this hath touched thy lips; and thine iniquity is taken away, and thy sin purged.
>
> (Vv. 6–7)

Psalm 60 again celebrates defeat over enemies. Oterunbi then prays for forgiveness. His prayer is punctuated by brief bursts of "tongues" as the Spirit takes control of his speech.[3] The collective recitation of the Lord's Prayer in English that follows, locates the C&S within the Christian tradition. But the concluding Yoruba prayer, known as the "Seal," betrays the performative Aladura gloss on "the power and the glory" of "Our Father." A prophet translated it as follows:

> The seal (aṣẹ) you have given to fire, that is how fire burns so bright; the seal (aṣẹ) you gave to the sun—it shines above everything on earth. Come and give this type of seal (aṣẹ) to our prayer . . .

It is now seven o'clock. We all stand as the next hymn, for "concentration," is announced. The familiar Anglican tune is sung in slow harmony: "*Mimọ, mimọ, mimọ, Olodumare*," "Holy, Holy, Holy, Lord God Almighty." The churchwarden carries clinking bottles of water up to the altar to be blessed, and so become imbued with the power generated throughout the service. Babies are rearranged on different laps, more members arrive. But the hymn deflects from these distractions, and encourages the incipient sweep of the Spirit through the congregation. As we then settle on the floor, some remain unaware that the hymn has ended, and continue to sing in tuneless Spirit. Prophet Korode, lean with fasting, remains standing before the altar platform, arms swinging, eyes shut.

Oterunbi, containing his own trance, continues to read chapter 4 of
the Book of Revelation, recalling the prophetic purpose of possession:

> After this I looked, and, behold, a door was opened in heaven: and the
> first voice which I heard was as it were of a trumpet talking with me;
> which said, Come up hither, and I will shew thee things which must be
> hereafter. And immediately I was in the spirit . . .
>
> (Rev. 4:1–2)

Next comes a long prayer for consecration, thick with references to
Biblical figures displaying spiritual strengths now similarly sought
from God, the origin of prophetic and performative power.[4]
Throughout the prayers and readings, the members murmur their
response: "*Mimọ*," "*Amin*," rubbing their hands together or raising
them in supplication. Individual involvement gives way once more to
collective participation in the thanksgiving hymn that follows. The
message is familiar: the strength and power (*agbara at'ipa*) that
comes from God, which now shows itself again in the bodies of the
members. By this time the hall is hot and full. Several sisters have
fallen in trance, the babble of their voices rising above the singing, and
drowning the beginning of the Yoruba thanksgiving prayer:

> . . . We thank God, because you appear to be in our midst, and are staying
> in the service with us . . . Since we started the service, we felt your
> presence in our midst. O Heavenly Father, accept our thanks . . .

Evangelist Ogundele voices the congregation's fears of illness,
infertility, miscarriage, madness, failure, and death dealt out by the
"hands of the enemies," thanking God for his protection. The Spirit
quietens once more; an elder sister's shouting gives way to a gentle
dove-like cooing, while six other women remain standing silently,
arms swinging, lost in the world of Spirit. A male elder is gently chanting in tongues; occasionally Ogundele's prayer is interrupted by his
own trembling, and brief syllables of glossolalia: "Wahalaa, whahalala."
Bayo Oterunbi dances from one foot to the other in front of the altar,
praying softly in trance.

But the "spiritual temperature," as members say, is set to rise with
the next part of the service: the three Yoruba spiritual songs. The harmony swells, the rhythms become more insistent with clapping and
stamping. A circle of men forms around the small, bespectacled figure
of Elder Ojo, a long-standing law student, swaying in the Spirit,
together with two other brothers. They dance frenziedly, their clapping
growing louder and faster. Ogundele drops to the floor, and remains

there, shaking. Around me, women are becoming possessed. Aladura
Sister Opayinka's broad body twists violently. Shrieking, she shunts backward over to the men's side of the hall, a sister on either side to support
her. She proves too much for them; falling flat on her back, she brings
them both toppling down with her in a heap on the floor. A couple of sisters
next to me smile, but others do not notice, hearing the hymns and their
own cries distantly through their trance, separated from reality by the
Spirit. The majority who remain unpossessed are nevertheless immersed
in the experience; swaying and dancing, they seem to embody C&S confidence in the vitality of the Spirit; they praise the creativity of God's
power, his protection against evil forces, victory over enemies, the place
of C&S among those iconic Biblical characters who won through with
the aid of the Almighty. They sing of God's vengeance against the faithless compared with the blessings for those who "hold fast."

The noise is now deafening; the cacophony of entranced voices
continues when the singing stops. "Jeeesuss!" a sister hisses; her
neighbor grabs her baby from her arms and wrestles to keep her slipping veil in place. Another woman is attempting to cover the legs of a
prostrate sister with her prayer gown. Two or three men sit with legs
outstretched, trembling and muttering. Prophet Aiyegbusi, attempting the first of three prayers for spiritual power, shouts above the cries
of women and the deeper groans of men. Elder Sister Adeoye follows,
a slight, lively, young woman, president of Esther Band. The third
prayerist, Prophet Ayorinde, a large, rotund man, who is soon to
return home as a qualified surveyor, repeats requests for spiritual
power, in English, at the top of his voice.

During the prayers, Prophet Korode remains in Spirit, but now he
regains consciousness sufficiently to join Prophet Ayorinde and
Evangelist Somefun in calming the congregation so that any vision
(*iran*) seen in the service can be revealed coherently. Spiritual Leader
Apostle Abidoye calls to the elders to pray for members shouting in
the Spirit so that they either speak out, be silent, or be taken out of the
hall. Korode goes over to quieten a shrieking sister, praying intensely
in Yoruba:[5]

> Father, come and open her spiritual mouth if there is any message you
> want to send through her. If not, then let her stop. May Satan (*Ẹmi
> Ẹṣu*) not cause obstacle in her way. Come and speak through the mouth
> of thy elected. Accept my prayer.

She falls to her knees, muttering quietly. Meanwhile, Tomori and
Ayorinde direct Lady Leader Awojobi, one of the most senior women

in the church, to restore order amongst the rest of the women. She goes first to Elder Sister Somefun, the evangelist's African-Caribbean wife, who is heavily pregnant; the Spirit had flung her to the floor some while ago. Two elder sisters are told to attend to another flailing sister; one places a Bible against her back and prays for "release." Immediately, she quietens. Another entranced young woman, not a regular member of the congregation, is speaking softly in Yoruba; a sister is directed to borrow my notebook to record what she is saying: "The peace of God like a flowing stream be with you." Ayorinde is also listening to Sister Elugbaiye, a recent spiritualist, who cannot make herself heard above the din. The official order of precedence in vision reporting during services and prayers gives first place to junior sisters like her, and concludes with the most senior male apostle. But this is adhered to more strictly in Sunday services, where the less experienced have a greater chance to speak. In Watchnight, the urgent Spirit of more established visioners is given space to override set order.

Korode now moves over to Aladura Brother Dele Faleti, a young photography student who had joined the C&S in London. Faleti is speaking in tongues; Korode, as a senior spiritualist, also uses glossolalia "to cut off his tongues." By using tongues himself, he later explained, he could both speak directly to God, and bring Faleti's voice under his spiritual control. He also hoped to avoid members' resentment or "embarrassment" at hearing Faleti brought to a halt. Although he judges Dele's spirit to be genuine, time is limited, and there is a priority in Ogundele, who was already an experienced visioner before coming to this country. Korode "respects his spirit very much." Ogundele still lies shaking on the floor, singing in tongues; Korode stands near him, answering him in chanted tongues, this time to "keep him going," Korode later explained, until the commotion subsides sufficiently for him to speak.

This takes some 20 minutes. Gradually, the clamor dies down, the sounds of Spirit fading to a murmur, as members settle on the floor. Abidoye and Tomori loll against the platform, some members are deep in private prayer, some reading the Bible or listening quietly, others are attending to their children, some seem to be asleep. Oterunbi's loud singing becomes soft, then ceases, as he bounces gently from foot to foot in front of the altar. Korode and Ogundele's chanting now takes center stage. With sweat pouring down his face, Korode strides firmly up and down, waving his staff with authority. The occasional thrust of his arms, and intermittent chuckling laugh betray a closeness to the Spirit, but he is in control both of himself and of Ogundele.

"Wasahaa, wasahalaa." When Korode stops singing, Ogundele picks up the chant, setting his syllables to a few repeated notes. He pauses; Korode, "inspired through the Spirit" and speaking in a loud, clear voice, "interprets" the meaning of the coded messages from God coming through Ogundele to the congregation. He then begins to chant once more, until the evangelist resumes. For 40 minutes this harmonious counterpoint continues. Sometimes their chanting voices overlap, sometimes they are sequential, weaving together in a web of sound. Never are they discordant, each taking over from the other in a melodic spiritual dialogue, smooth and carefully controlled. At one point, the message comes as a C&S chorus; at another, Korode temporarily departs from Ogundele's script to start up a song. The members' attention is immediately refocused through their participation, and the Spirit, never distant, ripples though the hall.

Ogundele has now been in Spirit for over two hours, so Korode decides to bring the exchange to a close. He stops and kneels with his head to the ground, the Spirit now shaking his body and forcing laughter from his throat. Ogundele immediately falls silent, remaining trembling on the floor until rising to his knees a few minutes later.

As if on cue, Bayo Oterunbi's voice now takes up the chant. Still slowly dancing in the Spirit, he starts to sing in tongues, interpreting the messages for himself in Yoruba. For half an hour he continues, but when, in the oblivion of the Spirit, his slow shifting dance takes him perilously close to the edge of the platform, Korode rises, touches him on the shoulder with his staff. "*El Satulah!*" (God who gives release). Using several other Holy Names of God "given" to him at that moment, he prays for the Spirit to leave Oterunbi. The effect is instantaneous.

Now it is the turn of the sisters, who should have been the first to speak, but although several women have shown signs of Spirit during the visions only one rises to her feet, reciting her messages in a quiet voice. Aladura Brother Babatunde, who follows, walks around the hall to reveal the visions he has seen. He is followed by Elder Adejare, a balding man in his forties who is the church organist and choirmaster. During Sunday services, if a prolonged chord obliterates the singing, this will be because Adejare has fallen into Spirit at the organ. Still lightly possessed now, he speaks jerkily in Yoruba, the sentences punctuated by throaty gasps.

A man at the back I do not recognize then stands up; the handbell cuts him off. "We do not generally allow you to say vision without first knowing or introducing yourself," says the Spiritual Leader in English. "We do not deny that your vision is correct, and you can write it down. May God be with you." A paper and pencil is passed to him.

Next comes Evangelist Somefun. He, too, parades around the church, waving his staff. The level of Somefun's dissociation is only ever betrayed by closed eyes, a few grunts, and rapid Yoruba speech; as trance recedes, he speaks in English. When he stops, he prostrates himself on the ground. Korode speaks next, in English, reporting those visions just shown to him that he considers important, that is, that require congregational prayer. Lastly comes Apostle Ayorinde, who gabbles his way through a string of visions, shouting loudly in English with accelerating speed. "Oh my God!" exclaims the Spiritual Leader quietly when Ayorinde finishes; everyone bursts out laughing. "Halleluiah!" calls out Abidoye, to show that no offence was meant.

Others will have seen visions, but, at 10.30 pm, time is running short. Visioners tell you that God will ensure that significant messages will "come out" through another speaker; if not, these will be noted and handed to the elders or to the individual concerned. As it is, all messages that clearly need congregational prayer have been written down by Elder Ajayi, the vision recorder. In English, he now summarizes these, and the members raise their voice together in spontaneous supplication. The remaining visions are quickly "covered in prayer" by an elder. Time also truncates the remaining program of hymns, psalms, and prayers, but not the sense of satisfaction with which the service ends. With Somefun at the handbell, laughing members dance and clap for the final hymn. Ojo is still in the middle of the stamping circle, jumping high off the ground, but now the Spirit has departed. A few minutes for silent prayer, the Grace said by Abidoye, and the clamor of the Spirit is converted into the sense of calm that members tell me they take with them into the night.

Possession in Popular and Ethnographic Discourse

The commotion of hissing and shrieking, the violent movement of white-robed bodies, the heat, the smell of incense mingled with sweat—all of this has no part in the staid worship of the Methodist or Anglican services that many members had left behind. But for C&S members themselves, possession is not remarkable; it is part of the routine, pragmatic work of the church. The Spirit was greeted in the Watchnight service not with gravity, but with enthusiasm, pleasure, and even—as with Abidoye's reaction to Ayorinde's fervor—with amusement. When joking together in private, groups of women, including spiritualists themselves, may mimic the trance behavior of male elders, jerking their arms forward with spasmodic grunts, shaking their shoulders and hissing through clenched teeth. Although "mockery of spiritualists" is condemned by vision and frowned upon officially, women find this imitation of men's possession-styles hilarious. Possession is fundamental to the practice of the C&S, but significance and humor are not mutually exclusive.

To the British gaze, however, especially two or three decades ago, possession seemed strange, both alien and exotic, and therefore fascinating. In 1971, the BBC filmed a documentary on the C&S.[6] The whole of Earlham Grove church was draped with lights and trailing cables, but the coverage of the service was highly selective. Ignoring the Leader's carefully constructed sermon, or the prosaic prayers, the cameras focused on the Spirit; they moved close to the female pews as sisters began to shake, and negotiated the crowded congregation for shots of elders contorted by possession. But by the time the visioners came to demonstrate the real significance of the Spirit through their revelations, the film crew had packed up and left. The C&S themselves were delighted with the program; their presence in British society had received some recognition, and their church accorded some publicity. But as there was no attempt to provide a social or ritual context for the proceedings or to explore the meaning, possession was fundamentally misrepresented. It was implicitly used to portray the otherness of African ritual, reinforcing stereotypes of erotic black physicality through images of trance.[7]

This interpretive slant by the media had, perhaps, as much to do with ignorance as bad faith. Trance is not intrinsic to mainstream Western culture, and images of possession are unfamiliar. In the 1970s, altered states of consciousness were popularly identified with minority, largely secular, activities, such as the counterculture's use of drugs, or abreactive forms of therapy treating an "alternative" clientele. Contemporary footage of audience reaction at rock concerts, such as "Beatlemania," suggests that teenage fans would also work themselves into a dissociative state. Religious trance was part of marginalized churches in terms of ethnicity, social class, and orthodox religion; it was practiced by West Indian Pentecostals[8] and white proletarian groups, such as the Elim Foursquare Gospel Church,[9] but was not found in the major denominations.

Even in anthropology, spirit possession had hitherto suffered a bad press. Since Oesterreich's 1930 study (*Possession, Demoniacal and Other, among Primitive Races, in Antiquity, the Middle Ages and in Modern Times*) it was often associated with abnormal psychology: as late as 1972, Weston La Barre could write: ". . . as individuals, prophets and shamans run the full gamut from self-convinced and sincere psychotics to epileptics and suggestible hysterics, and from calculating psychopaths . . . to plodding naifs only following the cultural ropes" (319). But the pathologizing of trance was dealt a serious blow by Erika Bourguignon, a pioneer in the anthropology of spirit possession, who showed that trance is found in the great majority of human societies, and that dissociation is a genetically inherited capability of all human beings (1968, 1976). Given these findings, it is remarkable that there have been far fewer efforts to account for the absence of trance in Western industrialized societies than to explain its common occurrence elsewhere. But at least the irrelevance of mental illness as an explanation of trance—even if it appears among conditions that possession may be used to treat—was endorsed by the majority of subsequent ethnographies of possession.[10] Indeed, the pendulum has now swung toward suggesting the positive, cathartic effects of dissociative states.[11]

The positive reevaluation of trance coincides with trends in contemporary British culture. The New Age interest in spiritual power, mentioned above, seeks to recapture capacities for direct spiritual experience seemingly suppressed by the rational positivism of Western discourse. This quest for individual expression has seeped into mainstream denominations, desperate for revitalization, through charismatic movements of the 1990s such as the "Toronto Blessing,"[12] or the Anglican "Nine O'clock Service."[13] In academia, the serious study of spirit possession was stimulated by Beattie and Middleton's classic collection *Spirit Mediumship and Society in Africa* (1969a), and Ioan Lewis's studies (1966, 1967) culminating in his *Ecstatic Religion* (1971), all of which suggest the positive sociological functions of possession. But it is in the last two decades that a burgeoning literature on the subject has produced a number of significant monographs, such as Lambek's work on Mayotte in the Comoro Islands (1981), Boddy's study of *zar* cults in northern Sudan (1989), and Danforth on the Anastenaria in Greece (1989), which move on from functional studies to focus on the meaning of possession for individuals and the community.

At the same time, charismatic Christianity, too, has begun to receive academic attention,[14] yet overviews of possession ethnography, such as those by Crapanzano (1977), Lambek (1989), and Boddy (1994), hardly mention it. It is here that the study of independent churches and indigenous Christianities such as the C&S can serve to bring together the understandings of possession derived from studies of both Christian and non-Christian communities, and take forward the analysis of the phenomenon of trance.

Trance, Possession, and the Spirit

If C&S trance is to be located in a cross-cultural perspective, it is essential to establish a common vocabulary. There are two main questions to be considered: first, whether we are comparing a similar psychobiological state in different contexts; and second, how far the meaning that Aladura assign to trance coincides with other cultural interpretations. The phenomenon of trance is but one of a range of altered states of consciousness, such as those reached in dreaming or hypnosis, achieved in meditation, or induced by drugs. As such, these assorted experiences are sometimes lumped together under the generic term trance.[15] Goodman (1988), however, has convincingly argued that "ecstasy," or ritual trance, as cross-culturally observed, is a particular psychophysical state involving distinct neurobiological patterns. Lambek (1989: 47), on the other hand, argues that there is no definable biological core of this state, which is thus as much a cultural phenomenon as the emic explanation for dissociation. But the accounts of trance in the great swathe of literature on spirit possession, despite culturally determined variations in behavior, agree so closely on fundamental characteristics that I would argue that we can safely use trance as an identifiable phenomenon for the purpose of comparative analysis of possession. This departs from Aladura terminology. C&S do not use the word trance (*iranran*) consistently: whereas some members equate trance with "being in Spirit," others identify this as a quiet, semiconscious state such as that reached during prayer, or in the liminal space that lies between sleep and wakefulness or even while day-dreaming at work. They tend to refer to dissociation by the force that is thought to have produced it: the Spirit. But since I wish to situate Aladura practice in the developing study of spirit possession, I shall follow anthropological convention.

Ritual trance, then, may be defined as a state of dissociation where normal consciousness is wholly or partially suspended and the subject experiences disengagement from ordinary reality. Access to external stimuli is restricted. We saw, for example, that those in Spirit during the Watchnight service were oblivious to disarray of veils and gowns (normally avoided for reasons both of modesty and ritual prescription). But despite deep dissociation, certain culturally determined stimuli can penetrate trance: in the service, the feel of the Bible placed on the back, a hand on the head, the noise of the clanging handbell, the opaque language of tongues, the sound of particular words and Biblical texts spoken in the confident voice of one recognized as spiritually superior—all evoked response from those in Spirit.[16] Cultural convention also determines the extent of amnesia experienced by the entranced; Aladura subjects may remember little or nothing of what took place during the episode. In C&S, this is also influenced by the depth of trance.[17] Prophet Korode once listened with intense interest to a recording of himself chanting in Spirit: "I had no idea it was like that," he said, having remembered nothing. Prophet Fakeye, who had interpreted Korode's tongues on this occasion and was not deeply in trance, explained that although he himself was "guided by the Spirit," he could recall everything, and was consciously using tongues to communicate with Korode's spirit: "This is where we have got to, so now what's next?"

The length of trance recorded in the literature on possession also varies greatly, from a few seconds to several days. Elders have often told me about prophets in Nigeria who remain in trance for up to a week, and bemoan the fact that in Britain pressures of work and study do not give them the opportunity to "sink deep into the Spirit," making prolonged trance of more than a few hours impracticable.[18] As the Watchnight showed, spiritualists may move in and out of trance, surfacing and submerging, throughout the service. Similarly, the ability to block out the normal physical responses of pain, hunger, thirst, and fatigue varies according to culture. In C&S it is often said that in Nigeria the true spirit is tested by holding a candle on the possessed person's skin or pricking them with a pin, for they will feel no pain. "You can knock yourself against the wall until the blood comes and feel nothing," said Prophet Ayorinde. I never saw anyone hurting themselves in Spirit, although spiritualists have told me of finding bruises on their body after trance. Even so, spiritualists claim to feel refreshed, and, if those who have been vigorously in Spirit for some time appear tired and drawn afterward, this may be because of a preparatory period of fasting, as Korode had undertaken, which can last for three days.

There is also a wide variation in the physical behavior recorded as accompanying trance, from wild, frenetic movement to quiescent ecstasy. In the C&S, as in the Watchnight, the bodily expression of dissociation ranges from a slight tensing or trembling, to repetitive swaying, swinging of the arms or dancing, to tottering around the room, to violent shaking and rolling on the floor. Vocally, trance in the C&S, as in many other cultures, may be accompanied by laughing, sobbing, grunting, hissing, muttering, shouting, screaming, or glossolalia, as well as coherent speech. Much has been written on the methods used to induce this dissociative behavior. Discounting the ritual use of drugs, forbidden by Aladura, emphasis is usually laid on the influence of driving rhythms in drumming, clapping, music, and singing. As Abidoye's exhortation and the rhythms of the service illustrate, C&S are aware of the effects of sensory stimuli. But this is mediated through both the meaning given to different rhythmic expressions and the mood of the individual. Ritual texts, as well as vibrating sound, equally seem to set off the Spirit. They serve to focus spiritual attention and encourage joy, significant in receptivity to trance. The sight and sound of people possessed also appear to send others into trance; spiritualists attempting to calm others say they have to be stronger not to catch their Spirit, which does indeed occur with the less experienced.

Once conditioned, the trance response may be triggered by a variety of stimuli; spiritualists confess to becoming possessed in inappropriate places when thinking about the Spirit—Dele Faleti when at work, Mrs. Adeoye on the bus. But in the capacity for trance, cultural expectation combines with psychological predisposition; however much they desire it, some members never fall in Spirit, while even prophets cannot always summon trance at will. In C&S exegesis the choice ultimately rests with God: "Some of our members climb around and make themselves unconscious by exhausting themselves, but the Spirit never comes" (Prophet Korode). And one senior elder sadly said: "I stayed on the mountain (in Nigeria) for three days and prayed and prayed, but I never saw anything at all."

The descent of the Holy Spirit in Aladura echoes aspects of possession-trance in many different contexts. But what emerges most clearly from studies covering a welter of diversity is one general but apparently paradoxical conclusion: every facet of behavior that seems at first sight to be so spontaneous and uncontrolled is in fact closely governed by cultural constraints. An event that seems to abandon order has structure imposed upon it and obeys the logic of the group ontology. Only when involuntarily experienced outside a social context, Goodman (1988) argues, does trance descend into madness, and is likely to be interpreted as demonic. The possibility of self-control and the regulation of trance, which is evident in the Watchnight, represents a sociological characteristic of dissociation in general. Despite its psychobiological base, trance is learned behavior; its onset, physical manifestations, duration, and depth are molded, although not rigidly determined, by the cultural context.[19] In the C&S, participation in group rituals and sessions of private prayer with spiritual superiors condition appropriate responses in the possessed. These are reinforced by attendance at meetings of the Visioners Band, which aims to provide training through supervised practice of prayer and revelation. Those who are not involved in this process, as the silenced man at Watchnight, are barred from publicly revealing the vision images they have seen in trance.

So also the understandings of the trance state, the explanations offered for dissociation, are embedded in cultural discourse. In certain societies, trance behavior is assigned a secular meaning,[20] but more often it is interpreted as possession by a spiritual being, to the invasion of the body by deities, ancestors, spirits, and other invisible entities. For Aladura, the incoming force is that of the power of the Holy Spirit, of *agbara Ẹmi Mimọ*. Not that they often use the English term possession; what they rather say is *o wa ninu Ẹmi*, "he, or she, is in Spirit." Spiritualists (*ẹlẹmi*) explain that when the Spirit "takes" or "catches" them and they "fall in Spirit," they are "taken away from ordinary life" into a spiritual dimension. Significantly, the Spirit is also said to "use" those it invades, which points to the pragmatic, prophetic purpose of C&S possession. Of this, more later, but here we can note that if no messages are forthcoming from one in Spirit, he or she will be silenced, like the shrieking sister in the service. The disappointed elder on the mountain was not just waiting for the Spirit—but to *see* visions sent from God.

The site of this interaction between God and his host is the *ẹmi*, the locus of personal spiritual power:

> When you fall in Spirit your own spirit is absorbed, covered with that Spirit, being one with the Spirit coming to you. So the two spirits are mixed together and joined up. The Holy Spirit and your spirit is joined to make one, which is what happened to the Apostles at Pentecost. They were having their own spirit before the other descended; it's like having reinforcements in the army, when there are some soldiers fighting and some reinforcements come to their aid . . . Then you feel the Spirit in your body, you feel it taking charge, and you can't control it with your ordinary mind.
>
> (Prophet Korode)

The Spirit therefore does not enter an empty domain; elders emphasize that for those with the gift of Spirit or visions, "the spirit is in you already":

> It's like a baby in the stomach. Sometimes it's sleeping and you think it's not there; sometimes it's rolling all about—but it's there all the time.
>
> (Prophet Fakeye)

In the C&S, therefore, "being in Spirit" subsumes both the state of dissociation, and the meaning that is assigned to the event; trance and possession are synonymous. In contrast to this, many spirit-possession practices do not necessarily equate possession with trance. Possession may be a long-term condition indicated by illness or misfortune; the manifestation of the spirit in the victim's body is only one sign of the invading spirit's presence. With the C&S however, the concept of possession outside trance is weakly elaborated. As we have seen, external causation is understood as an assault by another's power through medicine or witchcraft, as well as aggression of a malicious spirit, but is not described as possession. It may be suggested that certain individuals "have a bad spirit," that their own *emi* is under the influence of malign forces, such as *emere*, which could account for their instability and explain their troubling dreams. Occasionally direct invasion by an evil spirit is suspected; one prophet privately offered the opinion that Sister Opayinka's spirit was "dead," that is, that an evil influence caused her uncontrollable movement. But these assessments are often partial and idiosyncratic and do not amount to a dominant etiology of possession. Prayers are said for individuals or the church in general to "replace evil spirit with good," but there is no regular curative practice involving exorcism, no systematic attempt to accommodate and placate the molesting spirits, as in many cultures that diagnose possession outside trance. Possession in Aladura is not, therefore, associated with debilitation and disease. It does not focus on the victims of misfortune, but on the entrancement of visioners by God. For the C&S, possession is privileged as a positive and empowering practice.

The Sociology of the Spirit

If the Spirit is seen as an agent of empowerment, does C&S practice then endorse the connection between possession and deprivation, claimed by Ioan Lewis in his classic analyses of ecstasy (1966, 1971, 1986)? Mediumship may provide a strategy, argues Lewis, for individuals, especially women, to voice their demands for recognition or material benefits in order to offset ritual, political, or social deprivation. As to the sect, he maintains that possession is typically Practiced where the community concerned suffers fragmentation and oppression from the wider society.

Given the situation of Yoruba worker-students outlined in chapter 2, Lewis's conclusions on the connection between marginalization, relative deprivation, and possession at first sight seems appropriate. Viewed as black, working-class immigrants by hostile Londoners, disadvantaged in terms of housing and employment, the dispersed Yoruba community of the 1970s found themselves living on the edges of established society. But, as I also pointed out, there is an alternative interpretation of their situation, which is based on their own ambitions as an aspiring bourgeoisie; they are protagonists as well as victims. As Boddy rightly argues, the problem with Lewis's analysis is that it overlooks how people see themselves and their society (1989: 140, 278–279). C&S members did not think of themselves as an oppressed minority; rather than railing against injustice, or passively accepting an inferior lot, the church extolled their identity as "Stars of the World," their position as an upwardly mobile elite. Set apart from the working class by their educational pursuits and future professional status, they felt themselves superior both to other black immigrants in Britain and to those left behind in Nigeria. What they endure is a temporary stage in the transformation of their class position.

When it comes to the particular position of women, subsequent studies have built on Lewis's conclusions, seeing possession as a means to air female grievance,[21] or to realize female selfhood in an unindividuated society.[22] Clearly, Aladura did provide opportunities for women to take leading roles in the spiritual, as in the organizational, life of the church, and wield their influence through trance. But, as the Watchnight service showed, although the majority of spiritualists were women, most of the visioners speaking out in church were men. Usually, more women revealed visions than on this occasion, but C&S possession seems to mirror gender inequalities rather than redress them, for the public voice of the Spirit is largely male. Where it is female, although it adds to the woman's status and authority in the congregation, it cannot serve to express her individual needs and grievances. Although members look to their own dreams and visions as, in a brother's words, their "private oracles," the messages they reveal in public concern others, not the self.

This does not mean that church membership cannot provide compensatory status to offset underprivilege, and that the possession experience does not deflect from objective disadvantage. But deprivation, either on an individual level or that of worker-students as a group, cannot *explain* the phenomenon of possession in the C&S. Essential though Lewis's sociological analyses have been in stressing the significance of trance to the status of possessed persons or to the sociopolitical situation of groups that exhibit the phenomenon, the limitations of his approach come from the fact that he rarely deals directly with either the experience or the meaning of possession-trance per se. He declares himself wary of "styles of analysis [that] treat religion . . . as a thing in itself, with a life of its own, independently of the social framework which enfolds and moulds it" (1971: 15). But the missing link between the possessed and their environment is their own spiritual experience, the ritual meaning that they themselves attribute to their practice. It is from here that the sociology must start. What the C&S seek is empowerment, not just to deal with deprivation or cope with *anomie*, but to make the most of new opportunities. The Spirit, now in theory universally available, constitutes an empowering practice because it provides direct access to performative power for the ambitious and successful, as well as those demoralized by failure. What all require is efficacy to realise their long-term project, as well as to meet the disappointments of daily life.

Possession as Experienced Power

It is one thing to observe possession with an ethnographic eye, comparing cultural interpretations and expressions. It is quite another matter to experience it, to be swept into the maelstrom of physical sensation. The phenomenon of trance, despite diversity, has another common feature: it is an overwhelming physical and psychological event. In the C&S, trance is understood above all as *experienced* power. The metaphor the C&S most use to describe the sensation of possession-trance is that of the salient simile for spiritual power itself—electricity. They speak of electrical currents running through the body. The sensation may be brief: "It's like an electric shock, it really shakes you." If it continues, the spiritualist will enter deeper into dissociation, progressively becoming more deeply unconscious, "depending on the amount of power infused in you by the electricity."

Despite these explanations, spiritualists are not given to spontaneous descriptions of possession, because, I suspect, it is both too ordinary, being a standard part of C&S practice, and too special, as lying beyond words. When cautiously questioned, they speak of being taken out of their mundane selves, and into a different reality:

> You have to concentrate, put yourself into deep concentration, so you begin to lose awareness of your surroundings. Then gradually you feel the Spirit overpowering you, just like electric shock. You feel lighter than before; you are vibrating . . . it's like the engine of a car, once you start it, then you can leave it and it goes on by itself. Then you begin to see things, and begin to hear voices, just like a cinema, you see things with your inner eye . . . you retreat deeper and deeper into yourself to a place right inside you, where God is—where no one else can get at it—to that world inside you, that centre, where God dwells in you . . . The body flesh is just like a jacket: you can take it off and plunge yourself away from the body . . . this is how you are taken by the Spirit.
>
> (Prophet Ayorinde)

To be in Spirit is when the Holy Ghost actually comes down and is inside you. You can feel your head going far away and you know that it is coming. Then you feel that you are in a dream and no more a human being; the Spirit descends in such a way that you are out of your own mind and you won't know where you are . . .

(Prophet Korode)

Peter Korode, whose father and family are C&S stalwarts in Nigeria, first fell into Spirit at the age of seven. Although not conscious of the actual experience, he remembers the feeling "when I came out: fine and fresh, on top of the world, very light and healthy too." Now, as an experienced spiritualist, he still feels "great happiness" after being in Spirit, "the kind of happiness you feel after your wedding. You feel a different man, that you can only die a natural death and nothing else, because no hands can harm you."[23]

Other spiritualists speak of the elation preceding dissociation, followed by the post-possession tranquility that seeps through their body.

The longer you go on, the lighter you become, and when you finish, you feel on top of the world. Nothing and no one can trouble you; everything else is seeming just like a little thing. You don't feel tired or anything; you feel refreshed and strong.

(Elder Odunuga)

Even those who do not regularly become possessed may be familiar with this feeling. The Spiritual Leader, Apostle Abidoye, described his experience when he attended church in Kaduna, Nigeria, before coming to London:

I used to feel so big, my body would swell up. I would have to try and touch my feet because I felt that I would float up into the air. I *knew* myself to be big, and that no one could harm me; I was not afraid of anyone. The deeper you go, the more you know this, and this is the happiness in the K&S, the feeling of security.

The actual performance of possession speaks eloquently to the strength of the event. Yet when it comes to much of the ethnographic literature on possession, including studies of spiritual and Pentecostal churches, the reductive accounts of sociology and semantics seem to smother the raw power of physical sensation; rarely is there an attempt to capture the experience itself. Janice Boddy (1989), for example, sees a world of meaning in *zar* possession for wives and mothers, yet says almost nothing on the physicality of the experience, and what place this holds in women's lives.[24] Boddy is not alone: a plea made over 40 years ago by Bourguignon for more research on the subjective aspects of trance has largely gone unheeded.[25] Experience is sometimes ruled out of court completely: "Private feelings are of secondary importance in the sense that they are merely based on personal statements, the authenticity of which is not easy to verify" (Lee 1989: 252). How can you ever know what another is feeling, for how can private experience be captured and communicated in language?

The case for inclusion of subjective experiential data has largely been hived off into psychological anthropology.[26] Recently, however, there have been attempts to bring sensory data into the ethnographic mainstream. Emotion itself is no longer confined to the realm of individual psychology, divorced from sociocultural analysis.[27] Desjarlais (1992), Stoller (1989b, 1995), and others are exploring the sensory aspects of possession. With the C&S, sensitivity on personal emotional issues inhibited probing widely on this aspect of the Spirit, but the difficulties of delving into subjectivities should not lead us to ignore personal experience in analysis altogether. Here I am concerned with an epistemological proposition related to the significance of feeling and emotion: the existence of powerful sensation per se, whatever form this takes within the individual, has wider sociological implications. There is a critical difference between basing an analysis on the psychobiological details of sensory experience as articulated by the actor, and recognizing the collective significance of the fact that possession is a highly charged event. For the C&S, the vivid personal experience of possession has vital consequences in the confirmation and internalization of Aladura interpretations of trance as invasion by the Holy Spirit. It is hard to doubt the reality of God's power in the face of evidence that is directly experienced, internalized through sensation. When Paul *saw* the heavenly light and *heard* the voice of God, his conversion was instantaneous (Acts 22); Thomas's disbelief evaporated when he *touched* Jesus's wounded flesh (John 20: 24–29).

"Reality only exists for us in the facts of consciousness given by inner experience," wrote Wilhelm Dilthey, the nineteenth-century progenitor of interpretive anthropology (1976: 161; quoted in Bruner 1986: 4); his contemporary William James argued, "When . . . a positive intellectual content is associated with a faith-state it gets invincibly stamped in upon belief . . ." (1929: 496). A "faith-state" is an "enchantment," a "mystical seizure" (495). To be in Spirit, or to witness trance, is to enter such a faith-state, to participate in an archetypical self-authenticating event. C&S trance belongs to a class of inner experience, which through its very intensity proves its own existence, and legitimates Aladura theology.

In this, the Aladura are not unique. In any society, the fact that the drama of possession expresses a deeply satisfying individual experience carries crucial consequences beyond the personal. It lends weight to the meanings assigned to possession in the culture concerned, which are reiterated with every trance event. It invests emic interpretations with the certainty that comes from first-hand experience, transforming cultural "knowledge" into private conviction. With the C&S, it is the actual experience of a direct, physical encounter with the Spirit that legitimizes the whole ritual practice of the church, investing its etiologies with credibility, and confirming its understanding of *agbara*. In an age of competing religious affiliations, encroaching secularization, and increasing cynicism about the spiritual, every incidence of possession positively affirms the truth of C&S discourse, both to those who are entered by the Spirit and to those who witness the dramatic effect of the divine presence.

Clearly, for those outside the church, possession itself does not inevitably prove the power of God. Sceptical Yoruba in London laugh at those "shaking all about," accusing them either of submission to evil spirits, or of downright fraud. In members' accounts of conversions, a few have been "seized by the Spirit," and thus convinced, but more have been persuaded by the apparent effects of prophecy and prayer. But for those on the inside of belief, or hovering on the edge of commitment, possession throws the weight of emotion and sensory evidence behind the meaning assigned to trance by the Aladura.

This process is critical for the C&S, for the success of the pragmatic services it offers to its members depends on their acceptance of the Aladura cosmology. Divination through the visioning process, and the rectifying results of prayer, can only be interpreted as effective if the reality of spiritual forces affecting human lives and the superior power of God are fully accepted. It is this understanding of the universe that possession underscores. By dramatizing power, making it an event, bringing it into the center of Aladura practice, possession by the Holy Spirit becomes the lynchpin of Aladura discourse on *agbara*. "Discourses ... [are] ... practices that systematically form the objects of which they speak" (Foucault 1972: 49); the C&S exegesis of trance both recreates the meaning and the substance of spiritual power within the C&S.

Embodiment of the Principles of Power

As we have seen in chapter 5, the qualities of *agbara* are conveyed by the tropes of ritual texts. So the hymns, prayers, and visions of the

Watchnight remind the members of the properties of the spiritual power descending in the service. The texts allude to the hierarchical nature of power, the potential for spiritual supremacy or weakness. They imply the necessity of building up personal spiritual strength both to ward off malicious use of this morally neutral potency, and to gain mastery over positive spiritual forces, so realizing the full benefit of their revivifying, transformative capacity. But these characteristics of the Spirit are not only conveyed by verbal discourse, but are communicated, and so validated, through the C&S understanding of possession itself. The praxis of falling in Spirit, as the physical manifestation of divine energy, is understood to operate according to the same principles of power.

Vividly enacted through possession, for example, is a salient principle of power: its ability to transform. Through the descent of the Holy Spirit into their bodies, prophets and visioners are removed from the mundane, and transmuted into mediums of communication between God and his people. But the element of danger we saw as inherent in *agbara* is still present in possession: the Spirit may rebound on those weakened by sin or ritual impurity, treating them roughly, and violent but visionless Spirit may be sent to "cleanse" or even to "punish." Or, as was suggested in the case of Sister Opayinka, the Devil's agents (*emere*, witchcraft and evil spirits), may hijack the *ẹmi* of the vulnerable, those whose own spirit "is not up to date," in which case "they will be using you rather than you using them." For the whole theatre of possession acts out a central principle in the operation of *agbara*—that of hierarchy, where the weaker spirit must give way to the superior. The main statement of this conviction, underlining the metaphorical representations of power, is the event of possession itself, where personal power is swayed or overwhelmed by incoming forces. It is also central in the management of trance, as in the direction of a junior spiritualist or visioner by a spiritual superior. So Korode commented on the ability to calm spiritualists as evidenced in the service: "If you want to release someone who is in Spirit, you know the prayers and portions of the Bible to say, and the Holy Spirit will appear more [in you]. Spirit will bow down to Spirit, and if he is still mucking about, he will stop." The final expression of comparative power is the imperative to develop personal strength to meet, and use, the Spirit constructively.

These hierarchical principles of power embodied in the practice of possession were clearly articulated by Apostle Abidoye. He explained that power is inherited from the creation; this is what an individual is given at birth, but then it must be developed. To illustrate his point, he marked out levels on the wall with his finger. The top level, he said, is "that which Jesus came down to bestow on the disciples on the Day of Pentecost, the gift of the Spirit which rests on you when you are in Spirit." It is an "additional level," because it comes and goes. "You can't be in the top level all the time because you could go mad, and quite unable to enjoy life at all." But when it leaves you, you don't go down to the bottom again, but to the next level. This is the level gained by development: by reading the Bible, fasting, and praying. If you neglect this, your power will be reduced to the bottom level, which is what you have at birth. "Some people's power sinks right down again if they let it. And if the level is sinking, the space won't be empty, the Devil will come in, because he can have spiritualists on his own side."

This was not the only time I heard that "the Spirit can drive you mad," a further allusion to both to the forceful physicality of the Spirit and what the C&S understand as the potential danger of unregulated power. But control of spiritualists by their superiors during services— such as the curtailment of Faleti by Korode—must be carefully implemented so as not to be experienced as suppression; visioners also quote anecdotes of "brain-derailment" due to "squashing the Spirit in this country." They may therefore opt for self-control:

> When you are cut off, and they won't let you go on because of time, it is very terrible. The Spirit does not leave you completely, it is still there and can catch you again at any time. Now, when I feel the Spirit coming, sometimes I go out of the church and walk up and down, because if I pray, clap and concentrate the Spirit will definitely come.
>
> (Brother Fisayo)

The problem with this, say spiritualists, is that discouraging or neglect of the Spirit leads to its loss—a message we saw emphasized through vision-images of power. "The Spirit is like shorthand," said Sister Oguntulu. "If you leave it for three months without practice, it will slow down." Possession is a forceful physical experience, but as we saw in chapter 5, power—and hence trance—must be subject to human regulation.

Learning how to control embodied power and handle the Spirit is part of spiritual development, and is central to C&S praxis. As the Leader said, prayer, fasting, and Bible reading, gaining familiarity with C&S ritual procedures through involvement in the church, plus avoidance of polluting behavior are all said to contribute to the task of socializing the Spirit. A primary purpose of this self-control is the capacity to harness the Spirit to revelation—to "connect yourself to the telephone of Heaven." The relationship between trance and visioning is culturally established, for dissociation as a psychophysical phenomenon is not necessarily accompanied by visual imaging. In C&S practice too, possession may produce no mental pictures, for even an experienced visioner may be shaken by the Spirit without being "shown" anything on that occasion, while the novice may still be overwhelmed by the sensation of the Spirit in the body, or not yet have developed the ability to vision.[28] But members' comments, together with events during services (such as the Watchnight), show that flamboyant, flailing, but speechless exhibitions of Spirit gain little status in the church: "It's no good making noise and jumping all about if you can't tell us anything," said Prophet Korode. If causing a disturbance, an incoherent spiritualist will be silenced or removed from a service. Greater spiritual capital is earned by an individual's evident capacity to control his or her possession, in order to clearly interpret the visions seen, and so articulate the "messages" from God.

The same insistence on the instrumental use of power applies to glossolalia; tongues (*ebun ohun*) must be interpreted, otherwise, according to Prophet Fakeye, they "become ridiculous, and make the church a mess."[29] Loud, uncontrolled tongues are equated with negative, purposeless disorder through which "we will be losing our spiritual power." In personal prayer, tongues "are a private communication with God, when he wants to tell you a secret and give you so many mysteries that no one else can understand," but in public, tongues must serve a communal purpose.[30] So, as in the Watchnight, when a member is speaking in tongues without communicating meaning, a spiritualist of a higher grade will pray that he or she may speak out or be released. Alternatively, as occurred in the service, a translation will be made by the senior spiritualist, who will, it is claimed, simultaneously be shown the meaning in vision, hear the explanation through a voice, or know the message through inspired understanding (*imo*). Prophets maintain that tongues are not, in any case, inevitably associated with trance, and may be produced at will by those sufficiently experienced. In the chanted exchange between himself and Elder Ogundele, Prophet Korode, as the more powerful, could keep control of the episode by his own voluntarily produced tongues and so ensure that Ogundele's spiritual speech was interpreted. As Korode put it: "He's the source of the power, but you are acting as a control in the engine room."

When visioners are relaying messages received by themselves, they may be referring to visions they have seen earlier in the service (as was Babatunde and the sister who spoke at the Watchnight). If visioners are obviously in Spirit (like Oterunbi), they will be dealing with the messages of the moment. It is not always possible to distinguish the two modes, for a lightly possessed visioner (such as Adejare) may mix past and present revelations, slipping between consciousness and dissociation. Attitudes to manifest trance associated with revelation are ambiguous. On the one hand, prophets and visioners echo popular opinion that visions "go deeper" when delivered in Spirit, "because earthly somethings can't enter into it." "If you are not unconscious you can waive vision to your own advantage"—that is, consciously manipulate revelation. Falling in Spirit also underlines the message, investing it with divine authority: "If the Spirit wants me to do a certain job, it will force me to say it out. It will also make people pay attention to what I'm saying." But on the other hand, extravagant exhibitions of Spirit during revelation are considered by some elders to be incompatible with the *gravitas* appropriate to seniority. The last time Korode "fell down" was in 1960 in Nigeria; Ayorinde was the only prophet I ever saw prostrate on the floor. It is therefore not always possible to detect the depth of dissociation. With adepts, the presence of the Holy Spirit before or during revelation may be barely discernable, only a trembling of the hands, a fluttering of the lids betrays their state. For neither dissociation nor the visioning that may ensue necessarily entails involuntary movement:

> Falling in Spirit is not always about shaking, but consciousness. The Spirit is like air: when I breathe in and out there is no noise or force, it doesn't shake me, but it is still moving in me.
>
> (Evangelist Somefun)

Seasoned prophets can "sit down and see vision just like that." When talking with them, a slight pause in speech and/or a momentary closure of the eyes can signal that a casual revelation may follow in the conversation. This is not a sign of weak connection with the incoming Spirit, but of maturity and experience. "I am now so familiar with the Spirit; it becomes part of your body, and you can remain conscious enough to hold it," said Korode. A sudden shaking of a visioner's head during prayer may not be possession, but a conscious effort to resist complete dissociation in the interests of lucidity. At appropriate moments in the service, both Oterunbi and Korode were able to maintain their composure. The ebb and flow of Spirit among the congregation during the Watchnight service was therefore not fortuitous. It was regulated by ritual procedures—a dramatic portrayal of the necessity to control possession in the interests of the primary purpose of experiencing the Spirit in the C&S: the conveying messages from God.

The Spirit, Pentecostalism, and Indigenous Possession

The practice of possession and speaking in tongues would seem to line up churches such as the C&S with mainstream Pentecostalism of the period, where dissociation is also interpreted as possession by the Holy Spirit.[31] So Beckman (1975) characterizes Ghanaian independents (several of which identify themselves as spiritual churches and closely resemble Aladura) as Pentecostal. As for the C&S itself, Omoyajowo claims that it is "essentially a pentecostal movement" (1982: 98). Certainly, history would seem to confirm the connection. According to Peel, (1968: 63, 108) it was the arrival of Welsh missionaries from the Apostolic church in 1931 that brought the doctrine and practice of possession by the Holy Spirit into the center of a movement that already practised trance, visions, prayer, prophecy, and healing.

The way that trance is depicted in Pentecostal texts would also seem familiar to Aladura: so Hollenweger (1972: 334) describes "the Spirit and power," like "pulsating electricity," "a mighty divine power which penetrates into man" (330). But the relationship of humans to this possessing power is quite distinct in the two traditions. The Pentecostal reading is expressive, based on the New Testament bestowal of the Spirit on the apostles, who, as if "full of new wine," manifested their ecstasy in tongues (Acts 2:1–13). In ritual practice, the Holy Spirit is courted as a sign of a new, personal relationship with God. Despite theological debates, the Pentecostal consensus is that trance, and especially glossolalia, is evidence of baptism by the Holy Spirit and of a true conversion; its significance therefore lies mainly with the spiritual aspirations of individuals. In the era that the C&S was establishing its ritual practice in Britain, local Pentecostals were articulating a very different theology of the Spirit from Aladura. Lillie (1966: 39–40) asserted that tongues, besides showing that the speaker is selected by the Spirit, are an aid to private worship and the personal awareness of God; their interpretation is a separate and inessential gift. As for revelation, Michael Harper, a popular Pentecostal writer, did not mention visions in his *Walk in the Spirit* (1968). An Elim Pentecostal text on the Holy Spirit by Kingston (1964: 85–90) explicitly condemned solicited enquiry through prophecy.

For the C&S, as for Saint Peter at Pentecost (Acts 2:16–21), the key text on the Spirit comes from the Old Testament, and points to the instrumental uses of empowerment:

> And it shall come to pass afterward, that I will pour out my spirit upon all flesh; and your sons and your daughters shall prophesy, your old men shall dream dreams, your young men shall see visions: And also upon the servants and upon the handmaids in those days will I pour out my spirit.
>
> (Joel 2:28–29)

The C&S cite extensive Biblical precedents for this alliance of revelation and possession, such as that of Samuel, when "the Spirit of God came upon him, and he prophesied among them" (2 Sam. 10:10).[32] Through this invasion by the power of God, Samuel was "turned into another man" (v. 6), that is, transformed into a mouthpiece of God rather than into a reborn soul. The experience of the Old Testament prophets also provides a pattern for C&S practice: when "the spirit of God came upon" Balaam, he fell into trance, heard voices, and saw visions from the Almighty God (Num. 24:2–4). By associating the Spirit with revelation, Aladura adapted the practice of ecstasy in the light of Biblical precedent, and so related it to their problem-solving project.[33] Trance, therefore, became primarily valued as a first step toward revelation, not as an end in itself, for:

> Where there is no vision, the people perish.
>
> (Prov. 29:18)

Despite the historical connection with evangelicalism, there is as much resonance between C&S and indigenous, non-Christian possession, as between Aladura and orthodox Pentecostalism. Although we have little comparative material on trance in indigenous ritual, many of these conventions of Aladura possession also seem to characterize traditional practice. Invasion of devotees by their deities is a central aspect of most of the principal *orişa* cults. Through possession-trance the *olorişa* (priests) perform their primary role as mediators between visible and invisible worlds, and attempt to "transform gods from wild forces into socially useful powers" (Matory 1994: 219).

This process of taming the deities through embodying their power, so bringing them into the realm of the social, necessitates the same cultural management of trance as evidenced in C&S. Sango can seem wild and violent when he visits his priests, yet trance behavior is carefully

through his *orisa*, he himself did not possess; the High God did not convey messages directly through the means of trance. Significantly, neither did Orunmila, the deity charged with communication between God and humans. Orunmila spoke through the divinatory system of the Ifa oracle, in which the *babalawo*, the diviner-priest, read the fall of palm nuts, and advised his client without the help of trance.

What possession could do was to recast divination in a Biblical idiom, with vision and voices replacing palm nuts, and prophets assuming the *babalawo*'s role. The salient achievement of Aladura was the claim to establish direct communication with Olorun, fusing his power with divination, in a conjunction of information and power. Possession-prophecy and divinatory decision-making were welded into one system by the Holy Spirit. Possession-trance was ratified both by the central Christian text, the Bible, and by the familiar concept of invasion by invisible powers; Pentecost was read in the light of the indigenous discourses of power and divination, as well as of Old Testament prophecy.

As with many aspects of their practice, elders are quite aware of the potential parallels between the tradition they repudiate and Aladura. Whilst condemning them as "agents of the Devil, doing dirty and unnecessary things," Peter Korode once referred to Sango priests as prophets. The C&S like to think of possession by God as an abandonment of pagan practice. Yet the Holy Spirit, as experienced power, is also the critical means whereby a synthesis could be forged between Christianity and indigenous discourse to produce a practice legitimated by both. This synthesis is as significant for the C&S in London as for Aladura in Nigeria. Possession by spiritual power confirmed a distinct Yoruba identity, comfortably familiar, satisfying a cultural memory forged in feeling as much as exegesis. At the same time, the demonstrable communication with the Holy Spirit established Aladura Christian credentials and confirmed the validity of their faith.

Chapter 7

Revelation as Divinatory Practice

The purpose of Aladura possession, unlike traditional Pentecostalism, is to transmit messages from God about members' lives in order that appropriate prayer and remedial ritual action may be undertaken. An analysis of the subject matter of these revelations reflects the common circumstances of worker-students' lives. In essence, the purpose is to discover the play of unseen forces that lie behind these daily preoccupations, to divine the unseen causes that contribute to difficulties and dis-ease, or which might frustrate future plans. "Divination," Devisch says (1991: 12), "concerns the diagnosis and meaning of existential problems with a view to their management"—an apt description both of Ifa divining, and of Aladura visioning with which the C&S has replaced this indigenous oracle. Just as the analysis of falling in Spirit can be illuminated by the ethnography of possession, so I want to situate C&S practice in the context of African divinatory systems in order to contribute to an understanding of both.

The divinatory dimension of revelation is most clearly expressed in the personal consultation of a visioner by an individual client—a procedure to be examined in chapter 8. But the spontaneous offering of messages during services, group prayers, or unsolicited communication may also be analyzed in the same light. The difference between private and public revelation is one of context rather than process. Both, I shall argue, are part of the divinatory practice of the church.

Prophecy and Revelation in a C&S Service

Sunday, August 2, 1970: the annual Anniversary service of the Band of the Light of the Morning Star. Some 30 men fill the front three pews on the right of the church, their Band membership made conspicuous by capes and girdles of shiny green satin trimmed with gold braid. The service has gone well: a member's baby has been "named"; the Band has read its report, and handed over its donation of a large Bible. Further funds have been collected through the succession of church groups, who are called up to the altar platform to offer their thanksgiving: some 65 women and 70 men wearing prayer gowns sway from their pews at intervals, complemented by additional contingents from the 50 men and 70 women who sit at the rear of the church in ordinary dress. They crowd up to the front, dropping money or the specially printed donation envelopes into the plate, then kneel for a short prayer by the Spiritual Leader. So the white-robed members celebrate their collective identity: their association in Bands, their membership of grades and subgroups, their connection with the Cherubim and Seraphim, both as a contemporary congregation and as a specific Christian tradition. Together with those they have invited, or who have come out of curiosity or hope, they constitute a Yoruba community seeking to enlist the power of God.

But each one also carries an individual agenda of personal projects, of the problems that seem to obstruct their progress, and of the apprehension of unforeseen misfortunes, which could further complicate events. The testimonies by Band members during this service, as Adisa's speech recorded in chapter 3, bear witness to divine intervention in daily life: a woman has emerged safely from hospital, as promised by God; a member praises God that he is going home as foretold in vision, and is thankful that he does so as a member of C&S who has thus "learnt the full meaning of prayer"; an elder recounts how some time ago, after a service, a visioner had said that he saw him getting out of his car, while a prophet, to whom he had gone for "special prayer," revealed that he would be saved from an accident through his membership of the Morning Star. Band members had visited him to pray for protection; a near-accident experienced last week was a sign of the effect.

Although celebration of events in the church calendar, as on this occasion, reduces the time available, the hymn and prayer for the Holy Spirit, which follows the thanksgiving of the "invitees," prefaces the revelation of messages received during the course of the service. The first to speak is Elder Sister Oloko, a thin, single woman in her thirties, studying accountancy. She had been brought into the C&S by the first Spiritual Leader, who had been summoned to pray for her when she was discovered in her room having been possessed for several hours. Having now controlled the Spirit that she had dramatically exhibited earlier in the service, she speaks coherently in English.[1] Each of her messages opens with the conventional, "And as this one passes

away / was taken from me, the Lord said unto me / is telling us . . ."
Throughout each vision, communication with the Spirit is signaled
by: "and the Lord said . . .", and the authenticity of the message is
declared by the phrase, "before God and his people" inserted at intervals. She, like all visioners, ended each message or each section of a
longer vision with a phrase reiterating the importance of prayer: "May
the Lord hear our prayer"; "And may God continue to hear our
prayer"; "And may the Holy Spirit continue to be with us"; "May
God fulfil his promise for you"; "May God continue to open her
way"; "May God continue to protect us all." The necessity for ritual
action is thus an integral part of the vision: "Be fervent in prayer";
"Pray that this one will / will not come to pass—so said the Lord of
Hosts." In what follows, I have omitted much of this formulaic
repetition, as did the C&S translator of Yoruba visions. But they
constantly vouch for the divine origin and the veracity of the message,
and remind listeners that information is only one part of the visioning
process. Thus a frame of Aladura understanding is provided for every
prophecy.

Oloko begins with a couple of messages of general reassurance to
Prophet Fakeye, then addresses the whole congregation:[2]

> #1 Fear me, said the Lord of Hosts, for I am your Salvation. Repent, for
> I know my people and my people know me said the Lord of Hosts.
> I am coming to reign for a thousand years . . . That is the phrase I
> will say to you, that after this life will pass away, the Lord will reign
> forever. Be vigilant, pray, and trust only in me. I am your God and
> your Salvation. Fear nobody, because I will truly manifest myself to
> individuals, and I will show them that I am the Light, the Truth and
> the Way . . .

The next to speak is Lady Leader Awojobi. Now in her forties, she
arrived in England earlier than many members, and now works as a
machinist and as a representative for a mail order company. Both she
and her husband were long-standing members of a branch of the C&S
in Lagos—the ESO C&S, Morning Star. Her age and visionary experience had attracted a considerable personal clientele among members, which later was to form the nucleus of her own London branch
of the ESO. But at this date she was still one of the leading visioners
with the Earlham Grove congregation. Speaking in Yoruba, she starts
with a message to the celebrating Band:

> #2 The congregation should shout three Halleluiahs, because God is
> telling the President of the Morning Star that any sudden death

among the members, God has cast it away. Especially for three members of the Band, any trouble or impediment before them, God has cast it away today.

She indicates a brother seated toward the rear, telling him to pray that the "star" (life) before his child "be brightened, and that he may not go astray and be misled by life" (#3); the elders are instructed to do a prayer for protection for a "lady in the gallery with a scarf" (#4). After another seven such messages, she kneels; Prophet Aiyegbusi asks if further female visioners wish to speak; then calls on the brothers. The first to rise is Aladura Brother Babatunde, an energetic engineering student, stridently shouting in Yoruba. The content of his messages is reassuring and optimistic, and remains couched in general terms:

#5   I saw Jesus Christ stretching out his hand. He is telling us that whenever we are in trouble or difficulty, we should pray to him and he will conquer for us and rectify our troubles. Whenever there is trouble and misunderstanding in the church he will be with us.

#6   God is still promising that whosoever has the Holy Spirit, nobody will be able to assail or conquer them.

#7   God is telling us that we should bring the little children before him. He is promising that he will conquer for them; God is promising that the power of darkness will have not power over them.

After six more visions in this vein, Elder Ilelaboye takes over. Aiming to qualify in Business Studies and Company Secretaryship, Ilelaboye had joined the C&S and developed his spirit in London. Earlier in the service, he had been violently possessed, but now only quivers intermittently. His style is more discursive, revealing the image as well as his interpretation. The first vision (#8), delivered in English, confirms the presence of the Holy Spirit. He saw a "mighty cross descending to us." "God says this is the power given to us today . . ." He continues:

#9   . . . I saw that seven people are being brought into the church today; they are invited . . . and as the prayer was going on I saw that they were weeping. But God says that he is healing them today . . .

#10 . . . God says that you should continue to pray fervently, because I was taken to a very high hill, and I was shown so many things, before God and his people. When my eyes were opened, . . . I saw that most of us in this service today, their minds have gone astray. God is telling us that there are so many of us who are worshipping other gods, and we should desist from that one because most of their prayers have not been heard, because we have so many other things [i.e., alternatives to prayer]. God is warning all of us that it is better for you to follow him or to take another course. God is warning us seriously, that anybody who desists from his own way, may God not desist from us.

#11 ... I was taken to a river and I saw ... that most of us were washed in that river ... Most of us who have been sickly at heart, not just a question of ill-health, but in so many ways in which we are sickly, God said that he has pushed all this away.

#12 ... I was taken to a barren land, and there were so many people waiting there. And as these were waiting, I was demanding the understanding, and God is saying that most of us who have been demanding something of God, have not had anything to date. But God is telling us that this is the time to put our request before God, and by the time of next year, when we come to celebrate our anniversary, many of us will have cause to glorify God.

Four more visions, then Ilelaboye is brought to a halt by the bell. Time must be allowed for the prophets, whose messages are expected to be informed by their spiritual experience, and so be weighty and precise. The first is Prophet Aiyegbusi, a tall lean man who through his long C&S career and forceful personality had become a leading visioner in the church. He spoke clearly in English:

#13 In the name of the Father, the Son and the Holy Ghost; in the name of Jesus Christ. In the course of the service God said that we should help the leader of the service of today. For I saw something being removed from your neck, and the Lord said that from today I have cast this thing away.

#14 ... the Lord says that I should tell the Band of Morning Star that the Devil will try to conquer. There will be certain people elevated in your midst to a very high standard, ... but this will be in 1971, starting from April ... We should assist these people in prayer that this elevation that God has given to them that the Devil should not wage war against it, and that the rest people should pray so that the Lord will give them the strength to overcome it.

#15 The Lord said I should tell our Aladura Brother Adams, who is going home ... that you should not be afraid about your journey, because I saw an angel descending ... He will assist you greatly when you get back to Nigeria, and ... in countless ways he will perform wonders for you, and you should be sure that you are giving thanks before the Lord.

#16 And the Lord said that the promise he has made to your wife ... that there would be some change, because I could see that the Lord has opened her womb ... I saw that many, many things were being given today, and . . . that he has healed your wife today and that many things before her have been casted away.

#17   . . . The Lord said that I should tell our Sister Adeoye [president of Esther Band] . . . that you should have one prayer for your Band . . . with one day's fasting, and that you should pray to God that as God has established the Band . . . so that when this Band shall be coming forward to give thanks before the Lord at their next anniversary time, that many members will be full of joy before that day, because I saw the barren women among you—I saw that they came forward with children. May God fulfil his promise for you.

#18   . . . The Lord says that we should be very, very careful . . . mostly in our new church[3] because I saw that the first three services there, it was really a battlefield . . . Anybody who is going to lead the first three services in that church, they should be very fervent in prayer with fasting . . . because I saw that the Devil is definitely waiting there, but by the grace of God the Lord says that there is nothing to fear because everything is possible with him.

#19   . . . The Lord said that someone is preparing to travel to the countryside this very week . . . The Lord warns such a person to cancel such a journey, so that there will be no accident before such a person.

#20   . . . The Lord says that anybody who receives a letter that he should come home urgently, the Lord said that such a person should put it in prayer before he leaves, because I saw a letter being written . . . that he should come home very, very urgently, and the Lord said that such a letter is full of lies . . . They are just going to recall such a person so that they may block his way.

#21   . . . The Lord said that we should also pray for our accountancy people, . . . mostly in this fold, even in this country in particular, . . . because I saw . . . a heavy stone they are trying to put on them . . . that their examining body is trying to enact a very serious work to prevent them in future. The Lord said that we should pray for the people that our people may get out of this conclusion before their time is come . . .

#22   . . . So also I saw this country, the government of this country, they are saying that they are imposing laws . . . which will affect many immigrants in this country. But I saw that two of them are casted away . . . but that we should pray that the third one be casted away also . . .

#23   . . . The Lord said that I should tell you that why are you afraid about your people who you have left in Nigeria? The Lord said that he has promised before you left that he will protect them for you and that what you have come for in this country you will be successful . . .

#24   . . . To our Evangelist [man indicated]. There are three blessings before you, because I can see three fruits one before the other.

The Lord says that these were the three blessings before you, and that you should pray that these three blessings may reach you . . .

#25 . . . We should also pray for our people who have left for Nigeria . . . mostly Elder Sanni . . . for further protection over him, and that whatever he may be doing in his working place so that enemies and devils are not waging war over him.
May God be with you.

The next visioner to speak was Prophet Korode. Still exhibiting signs of the Spirit, he spoke in Yoruba, which he later translated for me as follows:

#26 I saw the Spirit like a dove descending on people. God is saying that he is sending not only power to see visions and dreams, but general Holy Spirit which he has descended on everyone in the service of today.

#27 For visitors who are pregnant. Anything which will bring them difficulty in the birth or bring them to operation or give them a deformed child, God has conquered it in the service of today.

#28 Warning against people who say that there is no prayer they can't do, because they are trying to force power not given to them. This changes prayer from yes to no when they come to the service, and causes delay and disruption in the church.

#29 Pray for protection over the owner of Oaklands [Congregational church from where the church was about to move], especially one of the top Executive committee members. Even if we do not know it, God has been protecting them while we have been here, especially that person. We must pray when we leave that church for the protection still to remain with them.

#30 To Elder Brother Asemota: I saw three pillars before him, representing his future progress. He should pray that none of these pillars should shake or be removed. God has conquered a sudden death before him, because I saw his name put on some type of medicine to destroy him, but God has cast this away and protected him.

#31 For the congregation. So many things are revealed to this fold through dream, but people feel reluctant and don't say it out. God is warning that they should speak out to the elders to avoid any trouble.

#32 We should be thanking God that he has made the anniversaries to be successful, because he had conquered misunderstanding in the Bands. We should pray that he will cast away future trouble of this kind from us.

#33 God is promising that all the presents before the Lord [from Morning Star], he will bless them for it. So no new member should dodge or fail to do his duty, and they also must not be proud for what they have done, so that the blessing may come to them.

#34 God has given us some special blessing today. We will be passing our exam more. Difficulties and failures will be diminished today in the fold.

#35 God is saying that we should not think that the Holy Spirit has left the fold. He is going to rebuild it, so the glory will increase more.

#36 Any time we do the opening of the new church we should pray for the people we are going to invite, for any misunderstanding, so that the power to bless the fold in London should reach them, and that God may continue to do his wonderful job with them.

#37 Any message going to the elders now, they should not neglect the vision, and they should do what it says. And when you are coming to the church, God is warning, do not take it as a joke. Nobody jokes with the word of God.

#38 Everyone should bring fruits to the next Holy Michael's Day to the fold, so that whatever you lay your hand upon it will be successful, and that the blessings should reach everyone. He will increase your blessings, especially the barren women in the fold, because God has promised that they will have child.

#39 One elder who has gone home—we should pray, because God is saying that there is a promotion before him, both spiritually and bodily. Because I saw a soldier with three large medals. Pray that it should come to him.

#40 Before Nigeria—I saw a big river, and everyone is trying to bridge the water . . . We must be prayerful—whatever may be the lack of expense [i.e., resources] not to be able to meet the financial problem—regardless of whether the war is finished or not, but because of bribery and corruption—so that people are leaving the country because lack of money or upset of mind, even without war, for God to conquer this.

At this point, Korode told me later, he saw a vision of wild eagles clawing the heads of members (#41). This, he said, was "bad spirits entering into people, especially women, to make them crazy." He "wouldn't say it out because it was too bad," and could alienate listeners who did not understand the safety ensured by a prayer for protection. He therefore "cut off" further vision to deal with it immediately himself.

". . . And to Elder Brother Asemota . . ." (#30). Asemota rises to his feet as do all male recipients of visions, whereas sisters slide from their seats onto their knees. When the vision points to a particular group within the church, or names the congregation as a whole, men and women in the relevant category acknowledge the message in these ways. Once the vision has ended, men will also kneel, in order to pray immediately over the information disclosed.

But a problem revealed is a problem shared: besides this private supplication, the weight of congregational prayer is put behind each request. As the visioners spoke, messages for church and congregation were noted in the vision record book for prayer and future reference; when the visioning was concluded, the Spiritual Leader summarized the significant messages, and invited the congregation to respond:

> We shall now pray for our Evangelist, that God's promise may be fulfilled; for all those who have left us, especially Elder Sanni that God may protect him, and for the other Elder that he may be elevated. Your prayer!

A cacophony of spontaneous sound ensues, as members raise their voices, arms aloft, until halted by the bell. The leader then "seals" the prayer:

> O prayer-hearing and prayer-receiving God [*Holy!*], come down and accept our prayer for our Elder Brother and his whole family that they may be protected. Let our members in Nigeria and elsewhere be blessed with promotion, through Jesus Christ our Lord [*Amen*].

Others mentioned in vision are called forward for prayer by a group of elders, as are "all the people now expecting, or in the course of expecting [hoping to become pregnant]." Time running short, he then asks the Pastor to:

> Pray for all the vision that came out, especially that the bad ones may be cast out, and the good ones to come out as quickly as possible.

Two further hymns, another prayer, and the service is over. But as the members leave their pews, some gather round individual visioners, kneeling for prayer, and hoping for personal revelation about their particular concerns.

The Conventions of C&S Revelation

When visioners say, "I was shown . . ." or "I saw . . .," they are referring to the images that appear in visioning. They describe these as static, "like a photograph," or moving "like the cinema or television," which they must then interpret. In understanding the meaning they may be aided by hearing voices or by a powerful intuition (*imọ*) about what they are shown. The images are often said to be dream-like; the similarity of the experience allows dreams (as in #26 and #31) to be included in the category of revelation.[4] Although vision and dream (*ala*) are distinguished, members would explain that "vision [i.e., revelation] includes dream," and that "deep and clear" "powerful" dreams are the first step on the way to visioning. The two experiences were subject to the same conventions of interpretation and reporting, and were seen as more similar than separate; when members referred to "night-vision" it was hard to know which they meant.

The significance of dreaming as communication between the natural and unseen worlds, as well as a means of personal divination, has a long Yoruba history.[5] But as Charsley argues (1992: 158–159), Christian practice may be stimulated as much by Biblical models as by "tradition." The key text is Joel 2:28–29, where God promises that the Spirit will speak not only through prophecy and visions. But elders pointed to further precedents for their practice:

> . . . hear now my words: if there be a prophet among you, I the Lord will make myself known in a vision, and will speak unto him in a dream.
> (Num. 12:6)[6]

Citing the experience of Joseph,[7] Daniel,[8] and Solomon,[9] members referred to dreams as their "private oracle," revealing the activity of hidden forces. (The expression "bad dreams" was a common metonym for the evil powers so revealed.) "Dreams are like a father's warning to his child," said one member. "They are more or less a searchlight for me."

As everyone can dream, all could share in the revelatory experience. But for the minority of members, their private oracle was also their vision:

> Through vision I will be shown everything that will happen, right up to a particular day. I know how it will start and what will happen next. . . . When it begins, if it is bad, I know how to prevent it, and if it's good, I can open the way to encourage it.
> (Prophet Fakeye)

Fakeye's comments underline the importance of acting on information received (on which more in chapter 8), hence the specific and general prayers for the circumstances revealed in vision at the end of the service. Messages concerning the church were noted in the vision record book, not only so that predictions could be verified by events, but so that ongoing prayer might be arranged by elders if specified in the vision or considered by senior elders to be necessary.

Although Sunday and Watchnight services were the chief public forums for visioning, messages from the Holy Spirit could be received by visioners at any moment: a sudden mental image in a mundane setting, in the ritual spaces in church or Band meetings, or during group prayer. But many of their messages were received while praying in private. If these concerned another, visioners might decide to undertake the necessary prayer themselves. But if they judged it necessary, they would speak to the person concerned, often by phone. After the obligatory greetings, these conversations would start, "I have a dream/vision for you"—not "*about* you," but "*for* you"; the reporter is a messenger of God, who is obliged to pass on the information received in order to respond to it with prayer. It was not only church gossip that kept telephone wires humming, but the reporting of dreams and visions, sustaining a web of constant communication throughout London—and beyond. The warnings and encouragements these calls contained were of the same nature as those made public in Sunday services.

Prophecies and Problems

Even if not individually addressed, what members hear from visions during services is evidence of divine awareness about their daily lives. Listening to the visioners' messages, they recognize their own situation in the circumstance of others. Accounts of members' experience in London outlined in chapter 2 show that most are familiar with the feeling of waiting in a "barren land" (#12), of being "sickly at heart" (#11) with no identifiable disease, but debilitated by anxiety and diffuse physical and psychological distress. From the standpoint of mainstream Western culture, where communication with the divine would seem an extraordinary event, C&S revelation appears surprisingly prosaic. But if visions seem repetitive, it is because the reality they reproduce is one of common experience. The routine referents of dreams and visions reflect the substance of members' hopes and apprehensions. This acts to confirm the authenticity of revelation as a way of knowing and influencing their world.

In order to gain a fuller picture of C&S revelation, I analyzed the visions recorded during a sample of eight services of four different types, giving a total of 678 messages.[10] What emerged was a picture of a largely personalized universe. Just as in the Band anniversary there is only a nod to political pressures on immigrants (#22) and the Nigerian civil war (#40), so only 2 percent of visions in the service sample concerns wars, famines and epidemics, or the accidents or assassinations of national leaders:

> #42   We should continue to pray for a country in Africa. Anything
> which may bring coup, for the Lord to avert it.
>
> (Brother Kolarin 18 January 1970)

> #43   There will be a war between two parts of a country. Pray that this
> will not come to pass in two or three years' time.
>
> (Evangelist Somefun 18 January 1970)

Messages concerning "a certain type of nation," or "a country which
I may not mention," refer to the civil war consuming Nigeria at the
time. Rules of vision reporting, reiterated in church announcements,
enjoined careful reference to Nigerian politics; the C&S as a largely
Yoruba church had to be seen to maintain an impartial stance.
Another reason given by visioners for euphemism, and for the emphasis
on preventive prayer always included in such messages, was that the
C&S should not lay itself open to charges of subversion—of planning
the disasters they exposed. Precision was only permitted where sabotage
was clearly unlikely: famine in central Africa, floods in Abeokuta. The
few visions concerning public affairs that were revealed in services
indicated the church's sense of political vulnerability, combined with a
conviction of their spiritual influence (#29).

But the subject of the vast majority of messages was far more
parochial: the personal concerns of Yoruba worker-students in
Britain. As would be expected, the return to Nigeria was a constant
theme (#20), found in 5 percent of the service sample visions.
Members are often advised to keep their travel plans private to avoid
jealousy, but many visions tell of safe journeys and positive prospects
(#15, 39):

> #44   To Brother Olaseinde . . . Provided you work well and walk
> rightly in his ways, the Lord is promising to raise you to a cer-
> tain post which you never dreamt of when you return to
> Nigeria. . . . It is not your qualifications which is going to give
> you this. Put it down in your diary, and if it is not so, it is not
> the God of Cherubim and Seraphim that is sending me the
> message.
>
> (Teacher Aiyegbusi 1 June 1969)

A further 5 percent of the service sample concerns exams and
qualifications (#21, 34). The message is again one of encouragement, together with the necessity for personal and ritual effort in

order to succeed:

> #45   To all those looking for success in their studies . . . you should
> read Psalm 119 over a cup of water and drink it . . . Wisdom and
> success will be yours.
> (Apostle Prophet Ayorinde 6 August 1970)

> #46   The Lord says that this year he is prepared to give you blessings,
> because I saw that those that are thirsty were given water to
> drink, and that those of you who have been searching for fish in
> water and cannot catch it. The Lord says that the meaning is that
> many of you have been working very hard and reading very hard
> towards the examinations, but when the results have come forth
> it has proved abortive. But the Lord says that this year will be a
> year of success for all of us.
> (Teacher Aiyegbusi 18 January 1970)

Encouragement is reinforced by practical advice: whether to attend a
particular college, whether to start, stop, or change a course:

> #47   The Lord says that there is someone who is planning to leave
> London for the countryside, and is waiting for admission to the
> university. But the Lord says that he should go to London,
> because the other one is a little bit dangerous.
> (Prophet Fakeye 13 April 1969)

The danger here is unspecified, but one of the potential obstacles is
the pervasive jealousy of others:

> #48   To the leader of the service: . . . you have been promoted in your
> education . . . but I saw people arguing with you, saying that
> why should you be promoted because your working in your
> school is not worthy of that post. . . . Be prayerful . . . and all
> those people backbiting that . . . you are not worthy of
> it . . . that this one will be cast away today.
> (Elder Sister Adeoye 7 November 1970)

A salient site for the assaults of enemies is the mind and body of
their victim: 9 percent of the sample visions deal with aspects of
health:

> #49   . . . He will help one of his junior brothers from abroad and pray
> that whatever may give him eye trouble for God to cast it away.
> Pray that the enemies may not make him to be blind.
> (Evangelist Somefun 9 January 1971; translation)

#50   [To a sister] God its telling her that whatever sickness she has
      before her, God has conquered this one today, and that before
      the end of the year, if she is fervent in prayer, she will come for-
      ward to give thanks to God.
                              (Brother Kolarin 15 June 1969; translation)

The ultimate fear of the outcome of illness, whether provoked by
enemies or not, is that of sudden death, a preoccupation that is
embedded in Yoruba culture. Divine promise of protection against
"anything which will lead us to the burial ground" or which "will
make us lament over" a member is reflected in the anniversary service
(#2, 30), and in 9 percent of the sample service visions:

#51   [To Teacher Aiyegbusi] Anything that may cause sudden death
      for his wife is cast away today. He must shout seven Halleluiahs
      before he leaves the church.
                              (Teacher Korode 3 August 1969; translation)

#52   To the Spiritual Leader: . . . the Lord is warning . . . the Spirit
      should have been taken from you, but God has promised that no
      one will die. But you must pray.
                              (Brother Kolarin 18 January 1970)

#53   To the Spiritual Leader: God is making another covenant with you,
      that we [i.e., the C&S] shall not lead anyone to the burial ground . . .
                              (Evangelist Olatunji 6 August 1970)

#54   . . . I saw many women, there was just continuous rushing blood,
      and God is warning our women that they must be very careful for
      anything that you may have committed by the hand of your own
      work that may result in sudden operation or sudden death . . .
                              (Apostle Prophet Ayorinde 6 August 1970)

The veiled allusion to abortion in Ayorinde's vision is only one
among many potential causes of death or disability to be averted by
vigilance and prayer:

#55   . . . One of the blessings which I am giving you today. Everyone
      who will be travelling by sky or ship will be arriving at their des-
      tination safely, for there will be no sudden death among you.
                              (Teacher Korode 1 February 1970; translation)

#56   To Mr Ajidagba . . . you must be very watchful against any
      accident, not a car accident, but bottle accident, which is very

dangerous. I saw people rushing out and plenty of blood . . . they were shouting "come and pray for him". . . . But this is a warning that if you do the prayer it won't happen. May God don't let it happen.

(Elder Sister Adeoye 7 November 1970)

Some 10 percent of the visions deal with such calamities. Members are warned of an array of accidents, from car and plane crashes, to collapsing walls in decaying accommodation. Visions anticipate theft perpetrated by friends or strangers, fire from upset paraffin heaters, assault from answering the door at night. Parents are warned about inattention to their children, and of the potential hazards of traffic, water, or unattended fires. Those intending to travel have their choice of dates confirmed or are advised that the trip is inauspicious (#19). The congregation is alerted to a variety of circumstances that could "implicate you," or "bring court case": lending money, carrying a parcel through customs or out of the office for another person, acting as a guarantor or witness without prior investigation.

Revelation also addresses a major area of anxiety: 7 percent of visions in the sample services deal with people and property left behind at home. Lost land will be regained; families will be protected (#23); no "bad letter" will be received. Reassurance about distressing news is especially welcome for those whose children remain in or who have returned to Nigeria. Visions reflect parents' concern:

#57   The Lord is warning to those that are sending their children home . . . either to Nigeria or to any part of the continent, not with the foster mother. You must be very careful nowadays that you won't just send your children to a lion, because I saw that immediately they landed, he or she was carried away [i.e., killed by malign power] . . .

(Teacher Aiyegbusi 18 May 1969)

God will also defeat those evildoers affecting reproduction (#16, 17, 27, 38). Some 6 percent of visions in the service sample deal with pregnancy, childbirth, and, above all, the promise of children to barren women:

#58   Three people are going for operation during the next week . . . You may read the Book of Ezekiel 37:1–10 in [i.e., over] water with three candles, and read this one three consecutive times and drink the water. I heard a voice say unto me that although the wombs are dead and dry, I can make them yield, and you will survive.

(Apostle Prophet Ayorinde 6 August 1970)

#59   To the Women's Committee . . . the next time you have your
meeting, you should pray with Psalm 24 . . . the Lord of Hosts is
descending upon you, especially those who are looking for children . . . the Lord will bless them with children, so you should-
n't doubt . . .

(Elder Sister Oloko 7 November 1970)

The present safety and future prospects of children are a familiar
theme in revelation (#3, 7):

#60   . . . I saw all of the children and they were given a handkerchief
today. The Lord says that we must pray over enemies or over bad
hand that will be obstructing our children this year, and God will
surely conquer for us.

(Teacher Korode 4 January 1970)

Some parents experienced particular anxiety if their child was living
with a "nanny":

#61   . . . I saw a women contemplating to remove her child from the
foster mother. The Lord is warning greatly that if such a women
should do this, she will regret it greatly . . . because I saw that
she did it herself [i.e., for her own reasons], not that the foster
mother neglected the child.

(Teacher Aiyegbusi 4 January 1970)

The reaction of many parents to a perceived problem with a foster
home was, indeed, to sweep the child away to place it with another
family. This vision—like many others—counseled caution. A major
moral theme in revelation was restraint, with warnings against retaliation, anger, and unconsidered action; visions advise patience and
tolerance in social and familial relationships:

#62   . . . although there is trouble and unpleasantness in the family
circle, God will help you. You are taken [i.e., thought to be] like
a leaf, which can pass away from a tree any time, but God . . . will
make you the rock in the family. So the Lord is warning you not
to quarrel with any of the family, or try to be aggressive . . .

(Prophet Aiyegbusi 6 August 1970; translation)

Some 8 percent of the sample revelations contain direct ethical
exhortation. Worldliness is mentioned, as is pride, but the C&S is less
concerned with a morality based on sin and guilt than with a code of

behavior consistent with social harmony and individual success. Visions condemn excessive smoking or drinking, promiscuity, adultery, rape, prostitution, and attempted suicide. They state that God will not countenance stealing, shoplifting, bribery, or dishonesty; they constantly warn against the effects of quarrelling, gossiping, and backbiting, both in the church community and in private lives, and they promise God's help in achieving harmony (#32, 36). Whilst some warnings are worded as abstract injunctions, others convey their message through dramatic vignettes (#56):

> #63   . . . I was shown a lady trying to find a knife to chop the man. Unfortunately they had to bring the man to hospital. They were trying to save the life of the man, but the man dies, and the lady is taken to court. God is warning, be careful. Women, if you are annoyed with your men, do not try to hold a knife or anything that can hurt him, so no man will die.
>
> (Evangelist Somefun 3 August 1969)

For God himself will punish injustice; he is aware of personal suffering, and revenge should be left to him:

> #64   The Lord said that he has been warning . . . that none of us will take our neighbours to the court . . . Somebody who is contemplating of doing such a thing . . . if he does it he will surely repent it. It is better for such a person to leave such a thing to God's hand because he has promised that he will fight for such a person.
>
> (Teacher Aiyegbusi 1 June 1969)

Perhaps Aiyegbusi "saw" the member in question here; perhaps later he spoke to him—but it is also possible that the recipient, if present during in the service, was never notified directly, or that "such as person" was never identified in the vision image. How is it, then, that members do not disparage revelation as being unspecific, but rather praise the visioners for homing in on their concerns?

Individual Agency in the Revelatory Process

It is not only in private consultation with visioners that members hope for guidance; all bring their problems with them to a service, laying their lives before the Spirit, and so solicit assistance by their very presence. Each ritual occasion, however large or small, thus assumes a divinatory dimension. In this public forum, a few messages are indeed quite precise, both in terms of the advice they offer, and as to whom the message is directed (#16, 48, 56). Take these examples of visions relating to the recipient's Nigerian identity:

> #65   And to our Brother [man indicated]. Somebody who is a chief in his family, or if not in his family, then his village chief. He should be prayerful for such a person. In future he may be called upon to replace such a person, but before he should take it up, he should put it in prayer . . .
>
> (Prophet Aiyegbusi 7 November 1970)

A woman is told to write to her husband in Nigeria advising him not to change jobs, even though the new offer is better paid. An elder is similarly told that there will be two posts waiting for him on his return; the one with a lower salary has better prospects. Prophet Korode once interrupted his spirited preaching to speak to the church Executive committee's legal adviser, a successful law student who was soon to return to Nigeria:

> #66   And as we were doing the Revival, before our Teacher Fadipe, before God and his people, God is telling you that the magistrate's work or Chief Justice will appeal to you in future . . . Do not hesitate, please, when it comes, because you will be better off there than practising law; so, though there may be struggle, this way is clear in the future. Pray that it may not miss you. So said the Lord of Hosts.
>
> (6 December 1970)

Yet, as the sequence from the Morning Star service illustrates, such precision is not the norm; few of the quoted visions are precise either on the identity of the relevant individuals or on the details of the circumstance affecting them. In the majority of messages delivered outside the context of a personal consultation or small prayer meeting, the intended recipient and/or the exact nature of their blessing or problem is left unspecified. In the sample of service visions, for example, 19 percent can be classified as unspecified blessing promised to the congregation at large. As with any text, written, spoken, or visual, the meaning is produced by the reader.[11] It is the role of the congregation members themselves that renders C&S prophetic practice relevant to its clientele, through the process of personally appropriating messages, then amplifying their meaning and interpreting the texts according to their own experience.

As in the Morning Star sequence and visions subsequently cited here, several of the messages in any body of visions will be directed to an indicated group or individual; "our accountancy people" (#21), "the congregation as a whole" (#31), "the Band of the Morning Star" (#2, 14), "the elders" (#37), "the leader of the service" (#13), "the lady with a blue headtie" or "in the gallery with a scarf" (#4), "Elder Brother Asemota" (#30), "Aladura Brother Adams" (#15).[12] But when this is not the case, the identity of the recipient will depend on *self-selection*, members appropriating messages through their personal connection with a particular category: "all those looking for provision of children," "our barren women" (#17), "three people who have applied for work in Nigeria," "all our examinees," "all those returning to Nigeria this year"; or, speaking to a potentially wider audience: "seven people . . . weeping" (#9), "all those facing obstacle," "those amongst you who have given up hope."[13] Often these messages addressed an unspecified individual: "a certain man," "anybody who receives a letter . . ." (#20), "someone here named Rebecca," "someone who is planning to leave London for the countryside" (#47), "someone who has been praying for provision of children"; "somebody in our midst today who has been sick for three years but was unable to tell another person that such a sickness has happened to me," "a man with pressure on his mind," "someone here today who has almost lost hope; he has failed one of his papers three times and is thinking that God has forgotten him . . ."[14] Given that the circumstances describing the intended recipient are often widely shared, any number of those present may take the message as their own:

> Three years ago I made a silent vow, and out of the three requests which I made from God, two have been fulfilled. And I was thinking that I would wait until the third one was fulfilled before I would come forward. But . . . in the last service there was a vision that someone had requested three things and he has had two, but he has not given thanks for them . . . My vow was open thanks and seven Halleluiahs, and that is why I am standing before you today.
>
> (Elder Brother Aluko.
> Thanksgiving Service 6 September 1970)

According to this testimony, not only had Brother Aluko assumed that a particular message was aimed at himself, but he had also provided the *amplification of meaning*, defining the three things by inserting his own projects into the inexplicit wording of the text. Revelation (as in visions #11, 12, 62) is full of such potential:

> #67   To our Elder Ajibode . . . Some people are coming before you, and are bringing some matter before you, but if the thing is not alright, do not support the thing.
>
> (Evangelist Somefun 7 November 1970)

An elder is told that "the battle you have been fighting since you were born will be ended two years hence"; a sister is assured that the "success you were looking forward to" will finally be realized; a lady leader is promised a positive outcome in "what you are praying for." "What you were planning to do, do not suspend it," a member is advised. Revelation repeatedly assures that "impediments," "obstacles," or "a certain trouble" will be removed.

When vouching for the value of revelation, it is common for members to cite visions where corroborating detail has in fact been provided by themselves. After the Morning Star anniversary service, a brother approached Teacher Korode to say that he had indeed received a letter from Elder Odunuga, recently returned to Nigeria, confirming the veracity of #39. In fact, Korode told me, it was Prophet Fakeye, also back home, who had been shown to him, although he had "not been instructed" to name the "soldier." The brother therefore had assigned identity to the elder, and fleshed out the vision according to his own experience.

Just as the nature of the problem can be amplified by the recipient, so the promised blessing can be understood in the light of an individual's particular concerns. A lack of precision about the benefits about to appear allows for the personal *interpretation of the text*, enabling the person receiving the vision to tailor the message to their own desires:[15]

#68 Before our Sister Odufona. I saw a shower of rain descending upon you . . . this is the shower of blessing . . . You have lost hope completely in your ways, but . . . the time is coming when [the Lord] will bless you even more than you expected. May the Lord hear our prayer.
(Prophet Aiyegbusi 6 August 1970)

#69 To our Elder Ajibode. The Lord says that he is your alpha and omega. He is the king who was with Jacob on his journey. I will descend in my power in your midst with my blessing. The Lord says that he didn't call you to come and worship him in vain, and he has come down to fulfil all your requests. God is continuing to promise you the upliftment of the Holy Spirit.
(Brother Babatunde 18 January 1970; translation)

Even where the imagery is more solid, the flexibility of interpretation allows for a range of referents to be inferred from one metaphor, to embrace a variety of meanings:

#70 To our brother at the back—yes, you, Sir. I saw you climbing a ladder, and half way up you became afraid. But the Lord says that, although the way has been difficult in the past three

months, you shouldn't be afraid. You must continue, because
today such a difficulty has been casted away.
(Prophet Aiyegbusi 7 November 1970; translation)

This vision was directed to a man in ordinary clothes seated at the rear
of the church, who might well have come to the C&S to seek assistance.
On the evidence of many personal accounts of revelatory relevance, it
is likely that instead of rejecting the message as unspecific, he would
be disposed to interpret the "ladder" in terms of his particular problem, and accept the encouragement contained in the message.

Occasionally, visioners publicly state the limits of their information:

#71 ...In 1971, in the USA...great things will happen in that
     country... but the thing which will happen was not shown to
     me...

#72 To the Spiritual Leader...[the Lord] will do that thing for
     him... but the thing he was asking I was not shown.
     (Evangelist Somefun 7 November 1970)

An "outsider" will offer financial help for the new church to avoid
borrowing—"but the particular man was not shown to me" (Teacher
Aiyegbusi 1 June 1969); a brother's life has been saved three times:
"one from motor accident, and the other two incidents were not
shown to me" (Prophet Aiyegbusi 6 August 1970). For those who
accept the premise of revelation, these admissions of ignorance do not
weaken the reliability of visioning. They rather reinforce the impression of a genuine dialogue between the visioner and the Holy Spirit.
Oblique allusion to information in place of full explanation in a message also conveys a sense of the visioner's experience rather than inep-
titude; visioners explain that their lack of specificity over identities
and details is often due only to discretion. Members familiar with
prophetic practice understand the C&S convention (which respects
the Yoruba concern for secrecy), whereby alarming or embarrassing
personal particulars should not be publicly divulged. It is for visioners to judge whether it is necessary to contact the recipient privately
to elaborate and advise, and messages may be left unexplained.
Often, therefore, once the visioner has decoded and articulated the
visual or aural image he/she has experienced to the extent that is possible or wise, the construction of meaning is completed by the recipient alone. Far from undermining the credibility of revelation, the
text is seen to constitute a personal, confidential communication
from the Spirit to the individual, to which even the visioner conveying the message may not have access; the connection between God

and the congregation is facilitated but not bounded by the visioners' prophetic function.

In practice, therefore, the space created for personal appropriation, amplification of meaning, and individual interpretation of a message greatly extends its potential significance to those present at its delivery. To members reasoning from within the logic of the visioning process, their role does not reduce their faith in revelation; the resulting accuracy confirms the power of visioning to illuminate past and present, and warn of what is to come.

There is little here that is exclusive to Aladura divination. Personal interpretation lies at the heart of contemporary prediction, be it by tarot, tealeaves, or tabloid astrology. Spirits convey metaphoric messages in séances,[16] and self-operated oracles such as *I Ching* depend on interpreting the texts to bear on the matter in hand.[17] So also with divination in many African contexts; C&S ideology, like that of the Ugandan Nyole, "ignores the very considerable role played by the consulter in divination" (Whyte 1991: 170); the arrival at a conclusion is, in practice, very much a "joint enterprise" (170) between diviner and the client, a process that I explore further in the context of private consultation in chapter 8.

Yet by no means are all messages amenable to personal appropriation, either because they contain no clues as to particular recipients, or because they are directed to members as a whole, in which case their content is such that it becomes impersonal. Of the 678 visions in the service sample, only 43 percent are directed to individuals (named or unnamed),[18] with a further 7 percent to members of a particular group. This leaves exactly half of all the visions directed to the collectivity: 8 percent to the C&S as an organization, and 42 percent to the congregation as a whole. If we exclude the two Watchnight services, where there is time to meet the expectation of individual visions, the percentage of general messages rises to 65 percent. Given the high proportion of visions that are not individually solicited and remain both general and unspecific, it must be considered whether it is helpful to identify revelation as divination, an ethnographic category most often understood as a process of seeking answers to definite enquiries.

Revelation and the Reproduction of Aladura Epistemology

Once, when I was discussing prophecy with Prophet Korode, I raised the question that had puzzled me for some time. If individuals relied so much on instruction from revelation, why was it that so many messages seemed to contain little specific information? Was it just a matter of repeatedly reassuring members of God's protective presence? Korode laughed, "You would say that! Look at Jeremiah!" As I had clearly missed the point, I did as he suggested. What I found was pages of visions and spoken messages from the Lord, a dialogue between God and the prophet. Voiced both directly in God's words and though the mediumship of Jeremiah, the visions speak of God's power, and his direction through trance, voice, and inspiration. They denounce other gods and goddesses together with their false prophecies. Replete with metaphor and image, they warn of the assaults and traps of enemies, evil people, false friends, of the dangers of conspiracy and gossip; they threaten the Lord's retribution, prescribe morality, promise his protection, and offer instruction and advice. The whole ontological background to Biblical prophecy—and to Aladura revelation—is there. In the same way, I realized, those visions that I had thought empty of precise content, without a specific target, were vital in reiterating Aladura assumptions, producing and reproducing the knowledge upon which C&S divination depended. For revelation acts beyond intentionality, not only to present the unseen forces behind a particular individual's predicament or encourage the congregation, but to invest ordinary experience with Aladura significance; to recast mundane events in a C&S frame of both meaning and action. The latitude allowed for individual agency in "hearing" messages, on which the relevance of visioning to individuals relies, rests on this reinforcement of Aladura convictions.

In previous chapters I have cited individual visions, divorced from the circumstance of their delivery, to illustrate aspects of this ontology, especially those concerned with the operation of spiritual power. I argued that as formal didactic structures are weak, these messages are critical in conveying C&S meanings to the congregation. But the significance of spirit-sanctioned messages in producing and reproducing Aladura knowledge cannot be appreciated by treating them merely as isolated units of meaning, whether in terms of their intended function as conveying specific information to individuals, or as educating members into the metaphoric conventions of the C&S. They must also be seen as parts of a sequence, as components of a repetitive body of revelations, which communicate Aladura interpretations of reality. Any series of visions, such as those to an individual client in private, serve to reinforce Aladura meanings in this way. But those delivered in a collective context such as the Morning Star anniversary, not only reach the converted, but also the visitors seated at the back of the church. These strangers heard themselves included in the promise of spiritual power (#8, 26), which was assured to the church (#35), as armor against all assault (#6), for the Devil and his "agents" were on the attack (#7, 14, 18, 25, 30, 48, 49, 51, 57, 60). For the benefit of listeners not yet socialized into Aladura practice, a revelation of enemy action judged to be too alarming will not be said aloud (#41). But the overriding subtext of revelation is to reinforce to adepts and newcomers alike what I have called "external causation," a personalized etiology that locates the cause of misfortune outside sufferers themselves. It may be secular authorities that are responsible for the "heavy stone" that prevents advancement (#21), and human deceit that ruins plans (#22), but this is far outweighed by the threats from invisible entities. The meaning of the "something" removed from the service leader's neck (#13) is the scourge of malign power, which only Aladura practice can counteract.

The assumption of enemy activity is so much a part of Aladura explanations that its presence may remain implied rather than specifically stated. Out of the 678 visions in the service sample, only 16 percent explicitly mentioned witchcraft, evil spirits, medicine, or employed those familiar euphemisms for spiritual attack: enemies or evil hands. But through the loaded language of battle and conquest, and the stipulation of preventive prayer, members are alerted to their movement (#5, 6, 27, 29, 32, 50), and advised to arm themselves with ritual defence.

At the same time, the search for protective empowerment from sources other than the C&S is condemned as both futile and dangerous (#10, 28), for it is only the Everlasting God who provides salvation from fear and aggression and opens the way to success (#1). The nature of these good or unwelcome prospects may be detailed (see later) or left unspecified (#4, 9, 13, 15, 24, 25), but dreams and visions are messages from God that must be taken seriously (#31, 37); they are revealed in order that action may be taken. Particular ritual requirements (#30, 38, 45, 58, 59), apart from fasting (#17, 18), are not always spelled out in public revelation; in the service sample of visions, only 13 percent offered ritual instruction. Directions are more detailed during personal consultation with visioners (see chapter 8). But in any context, recipients of messages kneel after their visions, and revelation will be concluded with prayer, for even if it not explicitly enjoined, action must always ensue. When members criticize "weak" or "watery" vision, this is not one that lacks personal detail, but one that provides no focus for prayer.

Messages may be couched in positive terms of blessings and assistance (#5, 24) or as more negative predictions of potential adversity (#20, 21); in the sample services, blessings slightly outweighed warnings by 53 to 47 percent. This partly depends on the personal style of the visioner: Prophet Ayorinde, with visions such as that of the rushing blood (#54), was known for his dramatic warnings, whereas Korode's messages tended to be more gentle and optimistic (#26–40). But the difference is one of emphasis rather than substance, for a constructive outcome in either case is contingent on prayer. As the Spirit warned an elder:

> #73 . . . when you listen to your vision . . . don't be so pleased with the good ones . . . because I saw that as soon as they tell you that you will be this and you will be that, you will be so happy, and rest your back on the chair . . . but you must be paying particular attention to the bad ones . . . and be fervent in prayer . . .
>
> (Elder Sister Adeoye 16 November 1969)

Over and over again, congregations hear this exhortation to prayer. Most do not have the time for Sunday School; many do not attend their Band meetings regularly. Yet by coming to the church and witnessing revelation, the basic tenets of C&S discourse seep into their understanding.

To introduce an analysis of divination with its epistemological under-pinnings is to invite charges of idealism[19] or of the essentialization of Aladura belief. But a subject-centered approach to divination such as that adopted by Devisch (1985: 77) must take seriously the promptings of the actors' own epistemology in the reproduction of the ritual sys-tem.[20] A synthesis of ritual ideology is therefore not incompatible with what Devisch calls a "praxeological" approach to the divinatory process, one that looks at the lived performance, participating individuals, and the ritual structure of the oracle itself. On the contrary, an integral part of such analysis is the "purposeful articulation of meaning" achieved by divination (77; also 1991), part of which consists of recasting the prob-lematic in metaphoric form. We do not have to rehearse the rationality debate[21] to argue that members take the ritual route as one strategy— amongst others—in the management of their daily life, because within the logic of their etiologies and understanding of effective action, it is eminently practical to do so.[22] Adisa, his co-religionists, and the prophets he consulted, perpetuate revelatory practice because, for them, it works. If indeed divination is to be understood as a "way of knowing" (Peek 1991b, c) or of "thought in action" as Zeitlyn (1987: 41) puts it, then it is essential to determine elements in this cognitive system that render it useful not only in providing existential order, but to enable the C&S to engage in the pragmatic activity of divining.

Meaning has performative value as well as ideological. The relationship between this belief, and the "explanation, prediction and control"
(Horton 1971: 94) it allows, is not unidirectional, for epistemology and
divinatory practice each constitute the other. Although Aladura
meanings are conveyed in all ritual discourse—prayers, sermons, testimonies, thanksgivings, and possession itself—they are primarily
legitimated through revelation from the Holy Spirit, where they are
articulated by what is taken as the authentic voice of God. It is visioning
that provides the essential epistemological frame for a Christian divinatory system. Messages delineate the unseen realities of spiritual
forces, and reinforce faith in the superior power of God, while C&S
interpretations of dream and vision images act to locate quotidian
experience within the Aladura cosmology, and thus render it amenable
to ritual action. It is the etiological discourse of revelation that carries
the weight of communicating the logic of C&S divination, for unlike
the traditional oracle Ifa in pre-Christian Yorubaland, Aladura was a
voluntary, minority movement. Although evolving from a shared
Yoruba understanding of spiritual power, the C&S Christian accommodation of indigenous concepts had to be conveyed. The ritual
representation of the Aladura worldview, through general as well as
specific messages, therefore becomes an integral part of their divinatory
practice, the background against which individuals could appropriate
and shape messages to address their own concerns.

Yoruba Divination and the C&S

They just go there to pass their exams—but we are not babalawo.
(Evangelist Opaleke)

Opaleke, like other elders, often referred to Ifa, the elaborate system
of divination central to indigenous ritual practice, at the heart of the
perpetual Yoruba quest to deploy spiritual power in everyday life.[23] As
we have seen, lesser deities did advise their worshippers through
trance, but the main route into the world of the unseen was the Ifa
oracle, which was not possession based. Faced with a problem, planning a project, contemplating a journey, struggling to find a cure for
sickness and misfortune—throughout life, an individual would turn to
Ifa or associated methods of divination for illumination on the interplay
of good and evil forces affecting both causation and cure. Ifa is seen as
the mouthpiece of the *orisa* Orunmila, Olorun's intermediary.
his diviner-priest, the *babalawo* (father of mysteries) is the earthly
representative of the deity.

The Ifa oracle itself is mechanical and textual; through the throw of palm nuts the diviner is directed to a particular *odu*, a segment from the body of Ifa lore he has stored in his memory. He then recites a number of associated verses (*ẹsẹ*), each one telling the tale of a mythic client's consultation: the problem, Ifa's advice, and the positive or negative outcome depending on the performance of the specified sacrifice. These legendary precedents are then taken to apply to the client's predicament.

The official discourse of the C&S roundly rejects all these indigenous oracles, including them in the "heathenish sacrifices and practices" that Moses Orimolade's original constitution "condemns and abhors" (ESO C&S 1930: 3). But contemporary Yoruba fiction bears witness to the fusion of *babalawo* and *woli* (prophet) in the popular imagination,[24] and C&S elders are quite aware of the continuity in divinatory praxis represented by their church:

> There used to be a kind of gap after the missionaries had preached against the *babalawo*, but then Aladura came in and filled that gap.
>
> (Pastor Tomori)

Peel's historical exploration of nineteenth-century encounters between "The Pastor and the Babalawo" (1990) charts this gap: what early evangelists offered was salvation from a future hell rather than from present misfortune or uncertainty. In place of Orunmila was a deity, Jesus, who, in one *babalawo*'s words, lacked the "power of hearing and answearing [*sic*]" (348), so leaving the worshipper immobilized in a limbo of spiritual silence. Perceptive missionaries recognized the existential disarray this caused to converts and the potential superficiality of an evangelization that separated the sacred and secular in daily life.[25] Orthodox Christian theology devoid of performative spiritual powers could neither incorporate oracular activity nor eliminate cultural practices that did.

That Aladura can compete for custom in the divinatory arena is recognized by elders as a promotional advantage. But they are also aware that this ontological continuity and the employment of spiritual power for divinatory purposes can result in confusion between C&S and tradition on the part of their followers as well as their critics (such as the contemporary Pentecostal Born-Agains described in chapter 9). So they avoid the word *divination* in reference to revelation. The word also carries unwelcome connotations of Biblical pre-Christian practice:

> There shall not be found among you any one that . . . useth divination, or an observer of times, or an enchanter, or a witch, or a charmer, or a consulter with familiar spirits, or a wizard, or a necromancer.

For all that do these things are an abomination unto the Lord . . .
(Deut. 18:10–12)[26]

Instead, however:

The Lord thy God will raise up unto thee a Prophet from the midst of thee . . .; unto him ye shall hearken.
(Deut. 18:15)

The replacement of oracles by revelation thus has sound Christian credentials; C&S members often refer to Old Testament prophets as visioners. Just as Aladura look for specific advice through personal appointments with visioners, so, they point out, Saul sought out Samuel,[27] Uzziah consulted Zechariah,[28] and David constantly "enquired of the Lord" as to what he should do.[29] Apart from these consultations, there is also ample Old and New Testament evidence for all the unsolicited information received through seeing visions and hearing voices: God so communicated with Abraham,[30] Samuel,[31] Elijah,[32] and Gad.[33] Jesus himself was guided by these means,[34] as were Peter[35] and Paul[36] amongst others.[37] As a sensible C&S pamphleteer remarked: "Vision is not a new thing among Christians; and it may mean a waste of the reader's time, otherwise there are more quotations" (Khita 1955: 3).

Allusion to these encounters in the didactic context of booklet and sermon, or in the ritual discourse of prayer and revelation, serves not only as an apology for C&S practice to the wider Christian community but also has an additional pragmatic function for the members themselves. As we saw in chapter 5, the principle of precedent is fundamental to the operation of spiritual power. Citation of case histories, as the C&S say, reassures the C&S that, as the Spirit once responded, so he will again, thus also reminding the Almighty of his obligation in this respect. The Bible is indeed the history against which divinatory visioning is set.

What visions are to prophetic divination, the ẹsẹ are to Ifa; historical precedent is embodied in the stories of the odu. As Abimbola comments: "The underlying philosophy of the whole system is that history repeats itself" (1965: 14). Ifa's verses cover the whole gamut of human experience in mythological form to serve as a guideline for present behavior. Unfortunately, however, there has been no full exploration of how the model matches with lived experience, for the recent interest in the oracle has not resulted in phenomenological accounts of Ifa in action. Apart from Akinnaso's illuminating description of a consultation in which he participated as both observer and protagonist (1995), we have no analysis of a session, or of a client's case history, as found in recent studies of African divination.[38] It has been the richness and complexity of the oral literature, and the evident skill and erudition of many *babalawo* that have most captured the European imagination,[39] and excited the justifiable pride of Yoruba scholars. The *odu* have been represented as "an inexhaustible repository of knowledge" (Makinde 1983: 117), encompassing the whole complexity of Yoruba philosophy, history, society, and natural environment, which still stands today as "the cornerstone of Yoruba Culture" (Abimbola 1994: 101; see also 1965: 14–15; 1977: 31–36). This privileging of the literary and philosophical significance of the *odu* has produced valuable records of texts, but left an ethnographic gap in the study of how Ifa operates as a popular oracle.

There is plenty of evidence, however, of its importance. In all its diverse regional forms, Ifa has been the most widespread of oracles throughout Western Nigeria, with variations enduring as a Yoruba legacy in Cuba, Brazil, and Trinidad, as well as flourishing among the contemporary Yoruba diaspora in the Americas.[40] Although declining in mid-twentieth-century Nigeria, Ifa was still consulted not only by the more "traditional" older generations, but also by Christians and Muslims. This raises the question of how far the historic oracles can answer to the experience of contemporary men and women, and in what way revelation provided a more vibrant alternative for a would-be elite of the period.

The Flexibility of Revelation

Although serving the same divinatory function, there is a fundamental difference between Ifa and revelation: C&S divination is not universally embedded in everyday life as Ifa once was; its messages do not embrace a hegemonic body of culturally created knowledge. For Aladura, it is the Bible that provides the cognitive context for visioning and the church that prescribes the morality, whereas understanding of the natural world is gained from secular and scientific sources. Yet the necessity of vision divination to underwrite its own epistemology is a strength rather than a weakness, affording it a flexibility and capacity for change essential for a modern oracle.

Modernity itself is critical, especially in the era leading up to Independence, and in the aftermath. In spite of cultural continuities, revelation exemplified freedom from tradition, and identification with a world religion that represented progress. Apart from the social significance of Christian practice, the cultural moorings of Ifa had come adrift for a would-be elite. The majority of members, both here and in Nigeria, were already Christians before joining C&S. Even if they were not practising members of an orthodox church, their education and employment had taken them into a culture more Christian than that of their parents; for them the theological underpinnings of the Ifa oracle were already lost. For literate Aladura, the Bible, more familiar than folklore, now replaced the *odu* as the mythological charter for divination. This, unlike the oral, specialist corpus of *ęsę*, was available to all consulters to study. In keeping with a historical period where individual achievement was high on the agenda, divination became more democratic.[41]

Yet despite the importance of the Bible to C&S visioning, detailed scriptural knowledge is not an essential prerequisite for either the vision-diviner or the client. The Bible is not primarily a divinatory text as are the Ifa verses; Biblical references are not essential to revelation. The system does not depend on cited precedents for its operation; inspiration frees divination from the text. Despite the reinforcing references to previous Biblical or C&S cases, it also cuts loose revelation from history. While Ifa is bound to a mythological past, dream and vision rely on immediate communication with the Spirit and have no temporal reference. Whereas many divination systems are concerned with the reconstruction of the past,[42] revelation, with its prophetic bias, looks toward the future, as do its London clients.

This is a future that will bring new problems, new circumstances unimagined by the *odu*, which can be encompassed by the creative freedom allowed to revelation. It must be said that the little evidence we have suggests that Ifa, although based on mechanical manipulation of objects and a literary canon, is much less rigid than might be assumed. Peel (1990) found the oracle adapting quite adequately to the needs of its nineteenth-century clients. Latitude in the interpretation of the *odu*, together with the potential for discarding outmoded elements in the *babalawo*'s oral repertoire whilst inventing others, allowed it to accommodate twentieth-century dilemmas.[43] Yet a system based on inspiration such as revelation has endless scope for innovation; there is no limit to the situations that can surface in vision-messages. Revelation can thus respond to the queries and anxieties of a varied clientele; it can speak to all sections of the elite, including worker-students in London. At the same time, as members reiterate, "vision has nothing to do with education"—and prophecy can also address rural workers and the urban poor. And not only Yoruba; although Ifa has traveled, it remains bound to its roots, divining by inference and analogy from Yoruba texts. Prophecy comes without this cultural load, and can, as it did in London, more easily cross ethnic as well as social boundaries. At root, the greater versatility of this Christian oracle comes from its basis in inspiration. Visioning is invested, as Ifa is not, with the explosive energy of a possessing Spirit, expressing a divination both more democratic and more flexible than the system it hoped to supersede. Historically, C&S offered mission converts an alternative to surreptitious consultations with the *babalawo*, by placing divination at the center of a Christian practice. In London, as both prophets and their clandestine Anglican clients told me, orthodox Yoruba Christians would make private appointments with visioners, covertly seeking the divinatory assistance that Aladura habitually enjoyed.

This page intentionally left blank

Chapter 8

"Practical Christianity": Revelation and the Power of Prayer

#1 [To a sister] I saw you holding two sticks . . . Two things will come before you very soon . . . and you will not know which one to choose . . . The Lord said that anything you are going to do . . . you should put it in prayer so that God may direct you . . .
(Prophet Aiyegbusi. Service 6 August 1970)

If the sister who received this message in a Sunday service was an experienced spiritualist, she might well seek this direction through her own dreams and visions. But she might also make a personal appointment with a visioner, who would put the problem before God in prayer, and offer revelatory guidance. Public revelation and private consultation are two aspects of the same process: visioners encourage their clients to involve themselves in the collective life of the church, but also, when they reveal messages in public, they may see that a member needs the personal attention of an experienced visioner. Often, clients refer themselves or introduce friends in trouble to a visioner. In these sessions, the visioner—like the *babalawo*—will not only reveal what God has to say about their problems, but also recommend prayer and ritual action. It is this deployment of spiritual power in private divinatory practice that forms the subject of this chapter.

A Prophet and His Clientele

Unfortunately, in exploring this aspect of visioning, there was a limit on what I could directly observe. The confidentiality expected of a prophet matches that demanded from other professional repositories of trust in contemporary Western society: the doctor, priest, counselor or therapist, aspects of whose roles the C&S visioner assumes. Given the Yoruba concern with secrecy, it would have been impossible to intrude on the unique opportunity for privacy provided by prayer, especially in the insecurity of diaspora conditions. Contrast this with most African divinatory practices that have fallen under an anthropological eye: the public nature of divination allows for ethnographic exploration both of the circumstances bringing the client to the diviner, and the aftermath of the consultation.

However, although lacking the client's side of the story, ample evidence on the practitioners' perspective emerged from diaries kept for me by three leading male visioners, each with an extensive clientele. While preserving the anonymity of their clients (to whom I have ascribed fictional names), they recorded basic information on those who sought their services, the nature of the problems presented, and the advice these clients then received.[1] I have focused on the fullest of these records, kept by a visioner I shall call Prophet Adesanya over a 17-month period from October 1969 to the end of February 1971, which is corroborated by the two shorter accounts. These records do not pretend to be complete. An exhausted visioner, up early for a day of work and college, attending to C&S business far into the night, has little time to recall every visit, or each detail of his dealings with his clientele. If he fell in Spirit during the consultation, many of his messages would be beyond recall. Quantitative data derived from the diary are therefore illustrative rather than definitive. Neither can a diary entry be taken as an exact record of the verbal exchange between visioner and client: it represents the synthesis of a consultation as recalled by one of the participants. Nevertheless, as texts that reveal something of what the visioner found significant in the event, the diaries convey a picture of private prophetic activity that underpins the public presentation of the C&S.

Prophet Adesanya had a wealth of experience in private consultations. Known for an awkward obstinacy, a quick temper, and a propensity, as he would admit, to "blast" those who annoyed him in the church, he had his critics. But as a prophet, he was, by and large, a popular and well-respected figure. Comparatively young, in his mid-thirties, his energy and his tenacity enabled him to combine employment in a Technical College workshop with pursuit of qualifications in Mechanical Engineering, plus a successful C&S career. With a wife, three children, and a problematic tenant, as well as obligations toward family at home, his own negotiation of all the aspects of his life had not been easy; he understood many of the situations presented to him for prayer from first-hand experience. But a sharp sense of humor, combined with a shrewd perception of people and their predicaments, allowed him to empathize with those who came to seek his help without becoming overwhelmed by their demands.

The chief impression conveyed by Adesanya's diaries and our subsequent discussions was the sheer weight of work that fell on a popular visioner in the C&S. There are 212 entries in his notebooks, representing an average of three to four appointments a week. These consultations could include more than one person; 42 percent of consultations were sought by couples (sometimes accompanied by children), 26 percent by groups of two or more (two or three couples, couples with a friend, or more rarely, a group of unrelated friends and/or co-members). Clients who came alone formed 32 percent of his recorded caseload. In all, the prophet noted that he had 371 requests for prayer during this period by 136 individuals: 61 men and 75 women. The great majority of clients are recorded as having made between one to five visits during this period. But some relied on the prophet to follow the intricacies of their lives more closely: 16 appear to have visited on 6 to 15 occasions, and a further 7 consulted the prophet up to 25 times or more.

One of the other diary keepers with a geographically more accessible house recorded an even higher level of consultations. But from my knowledge of their daily activities, both prophets had a considerably heavier schedule than they record. Besides this, none of the diaries includes all participation in further group prayers, which were held either in the church or in private houses to meet the needs of members. A prophet had to be ready at any time of day or night to respond to a phone call or a visit from a person asking for prayer. Often visioners— as other C&S elders and aladura brothers—would be driving around London far into the night. If this was by mini-cab, it would be at their own expense, for no charge could be made or gift received in respect of prayer.[2]

In addition to praying for those who so request, prophets and leading elders would also take on prayer and fasts on behalf of members unable to do this for themselves. This ritual surrogacy was necessary, elders claimed, as many who sought out C&S visioners for prayer were spiritually inexperienced. Only 28 percent of Adesanya's clientele who were members of C&S had already joined in Nigeria; the majority (72 percent) had become involved in England. But the prophet also attracted clients from outside the church: of the 127 clients noted by Adesanya, 37 percent never came to the church at all; these nonmembers accounted for 20 percent of the personal consultations recorded in the diary. Of the 81 clients who did attend the services, Adesanya reckoned that some 52 were regular members, participating in church life, while the remaining 29 came only occasionally. Therefore only a minority (41 percent) of the clients appearing in the record were active members of the C&S.

Even this distinction is hard to maintain; members could sustain different levels of activity at different times. The comings and goings of a couple in Adesanya's clientele are a case in point. Familiar with Aladura from Nigeria, Mr. and Mrs. Olawo had already visited Prophet Adesanya for prayer in the past. But the prophet had held back, sensing that they were involved in using "medicine," a suspicion that they later confirmed. But in the disaster that brought them to his door again, remedies from home had done nothing for them. Living on a council estate in Shepherd's Bush, they had been subject to constant racial harassment and were unable to secure police protection. When they contacted the prophet for help, he first advised them to come to the church on Sunday, because he felt that "they just go for vision when they have problem, like going to palm reading." This they did, but after the service Adesanya asked another prophet to pray for them, because, he said, he "did not want them to think that my vision came from what I knew." Adesanya did then start to see them regularly; the local authority offered them temporary refuge and took the children into care before they moved back into the block. The prophet records:

> #2 Domestic trouble against theft in their council flats that people come to threaten their life and break their windows . . . and the case has gone to the board of race relationship.
>
> Vision both in my house and when they come to the church that God really needs them now and their difficulty will soon be minimised through prayer.
>
> After a few days prayer everything is coming back to normal and after leaving their flats and leaving their children with the local council they [are] now settled back into their flat. Although the house was consecrated . . . before moving back.
>
> (11 October 1969)

Adesanya had helped them contact both the Race Relations Board and the Housing Department. He had also prayed over the selection of one of three alternative flats the Olawos were offered. However, they ignored the vision's recommendation and opted for one of the others, despite the drawbacks specified by the Spirit. This then fell through, by which time the third flat was already taken. The prophet agreed to protect their old accommodation with prayer, but warned them to respect revelation in future.

Although often omitted from the diary, Mr. Olawo was, the prophet said, coming "almost every day" and praying with Adesanya from 10.30 p.m. until midnight. If his wife accompanied him, the prophet would also spend time discussing her problems with her. An additional worry they brought to him at the end of 1969 was their children's health:

> #3 After going through the housing problem one of the children is sick. Warning in vision not to send for native medicine from home to help the kids as this will cause obstacle in the kids' life.
>
> (8 November 1969)

The prophet would also call at their house to bless water and see how things were progressing:

> #4 . . . there was vision that the children would be OK through prayer and that they should keep on consecrating their house at all times not to wait until there is trouble, so that trouble can be averted from them.
>
> (26 November 1969)

Visions continued to confirm the children's recovery:

> #5 . . . but they should move nearer to God and be prayerful with certain psalms. Also to beware of friends who pretend to help them but instead they try to destroy them morally and spiritually.
>
> (7 December 1969)

By February 1970, Adesanya's records show him still pressing for the couple's closer identification with the church:

> #6 Vision to pray more and if possible get their prayer gowns.
>
> (20 February 1970)

> #7 . . . to consecrate a place for prayer in their house, etc.
>
> (28 February 1970)

But by then Mrs. Olawo was ill, and the marriage was running into trouble. At one point Mrs. Olawo had left her husband; Adesanya persuaded her to return, but the relationship continued to deteriorate:

> #8 To collect the water already blessed for his wife who is sick, and general protection and healing for them.

Vision that he should be grateful to God because if they have separated earlier when they had quarrel, people will formulate that the sickness would be that he caused it by poisoning his wife and from there wage secret war against him. God will actually help them to cast away all the sickness in the family and that nobody would die. This year would be the end of their difficulty God promises.

(24 March 1970)

Some things did improve:

#9   Prayer for more protection as the wife has just been completely healed from chest trouble.

(3 April 1970)

But the children were not well, and neither was the marriage:

#10   The child is sick overnight.
        After the prayer it was revealed that the child will be OK and [they] should not be afraid over the sickness. They should bath the child with consecrated water.
        They should not wait till any of them is sick before holding prayers in their house because of some other blessings too [i.e., that would come their way if they prayed regularly].
        If they want to help a relative they should be careful so that this will not be in vain and cause misunderstand[ing] even between the couple. In fact the last paragraph was what they need my advise for after the prayer, as they have already quarrelled over a relative as such.

(6 May 1970)

As well as illuminating past plans and present difficulties, Adesanya's prophecies also revealed the potential in the family's future if they could persevere:

#11   Prayer of thanksgiving and more healing.
        Vision that they should beware of petty quarrel that leads to big ones as this is the weapon of their foe to destroy them after which serious sickness follows in all cases.
        Several times they have planned to attempt and do good thing but half way the plan is change[d] from good to bad. This is the work of the Devil.
        They should be praying at bed time with Psalm 46. They have to thank God that the sickness could again be [i.e., has not been] transferred to the youngest son. They should look after this child

and he will be a glorious and clever student. Law and medicine
will appeal to him in future. The early part of law is bright but not
the end so by this time they should put both in prayer. They should
beware of taking gifts from friend and relatives as this was planned
to be poisoned. The only way to overcome the long-ranging difficulty before them is to move nearer to God, etc.

(27 July 1970)

Adesanya's support continued for several months, dealing with a
succession of issues, including financial difficulties and employment.
But by mid-1971, even though the Olawos did attend the church,
they no longer formed part of the prophet's clientele. For his part, he
felt that there was no purpose in continued prayer unless they were
prepared to take more responsibility in resolving their own problems.

Revelation and Daily Life

When Adesanya ends his notes with "etc." (#7, 11, 12, 15, 24), this
often indicates that he does not recall the remaining messages. But
et cetera can also be read as alluding to the common occurrence of the
situations reflected in his visions. Over time, families such as the
Olawos are likely to present a typical range of problems; the pages of
the notebooks are filled with requests for prayer around the regular
preoccupations of worker-students' lives that we have already seen surface during Sunday services: health, education, employment, housing,
lawsuits, marriage, childcare, family in Nigeria, future prospects, fear of
sudden death. All these reappear in private consultations. Mr Olawo
could not afford to study and was employed in fulltime factory work,
but it is anxiety over gaining qualifications that predominates for men.
The diary captures their despair, and the hope that revelation offers:

#12　General Prayer [for a couple] including failing his exam in one
　　　subject for five or six times.
　　　　　Vision: . . . The slippery step he was climbing regarding the exam,
　　　God will let him pass it by flying instead of climbing the steps, etc.
(23 July 1970)

For women, the concern that predominates in the diary is the
whole anxious question of reproduction. This is also believed to be an
area especially susceptible to the operation of unseen powers, a threat
that lurks throughout the entries. Such a case was that of Sister
Adewumi. Although she already had one daughter in Nigeria,
her failure to conceive again in England was provoking the hostility

of her in-laws:

> #13 Prayer over her dream which came through [i.e., which she had,
> arising] from the letter [she had] found written to her husband
> from home [from his family pressing him to leave her]. For God
> to change the mind of her oppressors and protect her so that she
> may not lose her husband.
>
> Vision that it is true that her [husband's] family foe are wag-
> ing the war against her and that the native medicine that was
> prepared for her by her [own] family was interchanging from
> good to bad [i.e., although intended to help her, was having the
> opposite effect because of this]. She should be grateful to God
> because if not because she is in London the opportunity of
> Cherubim and Seraphim might not be opened to her and she
> might have died through her oppressor. Not only her that was
> troubled but also her husband and this is why he hardly have any
> satisfaction in what he does or what belongs to him. She should not
> worry that she has not got a male child as God will actually provide
> this for her when the time comes and return through [i.e., true]
> love between both of them and that she should learn how to fast
> and pray. Through this things will work her way.
>
> (15 July 1970)

"Foe," "oppressor," "enemy"; "waging war" with "poison" or
"native medicine": these menaces appear as regularly in the
prophet's private revelations as in Sunday services. The influence of
malign power constantly crops up in the Olawos's consultations
(i.e., #3, 5, 8, 11) and emerges as a principal factor in many clients'
experience. The salience of what I have called "external causation"
could be read as a metaphorical statement of insecurity and a sense
of threat. But on a cognitive level this understanding of the world
serves to *personalise experience*. This individualized approach
to events and life situations essentially deals with the currency of
relationships—whether social, physical, psychological, or spiritual. It
is with this dynamic view of causation that revelation deals. Many
predicaments are represented in vision as *social* relationships:
between real and false friends (#5, 11, 25), between an individual
and potential critics (#8), between neighbor and neighbor, landlord
and tenant (#16), applicant and employer (#15), husband and wife,
affines and kin (#8, 11, 13, 10, 20, 24). Racism, for example, is represented, as it is indeed experienced, as individual prejudice rather
than in institutional terms (#17); candidates are often instructed to
pray for "mercy on the exam"—to predispose the examiner in their
favor (#24, 26). In matters of health it is the *physical* relationship of

the body to the environment, both material and unseen (#9, 11). Then there is the *psychological* element: the relationship between mind and body, mental state and affective function (#29). Behind all these are *spiritual* relationships: between the person and their life force, the individual and God (#5, 11).

Recasting experience into the idiom of relationships renders it susceptible to the exercise of unseen power. According to its operating principles (chapter 5), the impact of power is exchanged between animate or inanimate objects, not between people/objects and depersonalized events. What C&S members therefore seek is *empowerment*, reinforcing their impact on others. Here we return to another theme of chapter 5: power as a principle of general efficacy, the dynamic energy behind the relationship of individuals to their world. Although a client may present a single problem to a visioner, what is sought is empowerment of the whole person, together with those physical, psychological, or spiritual parts that appear to be most at risk. "Well, any time anybody dies suddenly, they say it's heart-attack," once commented the prophet, "but why can't the heart fight back when anything attacks it?" A number of clients are recorded in the diary as simply seeking "development of Holy Spirit." This may refer to progress in visioning and dreaming, but it is also because personal power, proceeding from the Spirit, is the prerequisite shield against negative influences and the foundation for a proactive approach to problematic relationships.

It is this personalization of experience coupled with the search for empowerment that enables clients to feel that they can have a handle on their problems through C&S divination. Yoruba worker-students could have little real impact on their social and economic circumstances, beyond political action, in which they were reluctant to engage. But through revelation and prayer, the alien, impersonal environment of migrants (both in Nigeria and Britain) was re-presented, and the individual rendered more effective to intervene.

Divination and Decision-Taking

But Adesanya's clients were not merely victims of circumstance or of evil spirits. They were ambitious individuals carving out careers. This involved taking decisions, making choices. Since vision can reveal the play of powers affecting projects, and disclose circumstances susceptible to prayer, visions encourage the spiritual investigation of future plans (#11):

> #14   Any job she will attempt to do in Nigeria should be put forward
> in prayer before undertaking such a problem . . .
>
> (10 March 1970)

#15   . . . Wherever he would work at home he should try and put it
      before the Lord not to mischoose, etc.

                                                        (1 June 1970)

#16   God will so mercy at the end [i.e., this project will eventually suc-
      ceed], but from now upward he should try and put his intention
      before the Lord before starting and not after . . .

                                                     (2 November 1990)

One convention in consulting visioners over decisions is to submit
unspecified alternatives to the Spirit in the form of numerals—the
"two sticks" in Aiyegbusi's vision (#1). This is reminiscent both of the
indigenous procedure in which two *ibo* (lots) were used to provide
positive or negative answers to specific questions, and the submission
of names or numbers to the Ifa oracle. The selected number, Adesanya
explained, then appears in vision as marked on a board, ticked on a
sheet, colored red, or is heard spoken by a voice. The reasons for
rejection or selection, or the ritual or other preconditions for a particular choice may also be revealed. Adesanya was offered numbers
representing children's foster homes and schools, colleges and training courses for their parents, for the exam paper that needed most
work, or for unspecified intentions.

Many of the decisions submitted to the Spirit concerned areas in which
the Yoruba student in London would have little previous experience on
which to draw. Alternatively, they dealt with familiar life events around
which cultural norms were changing. Here divination negotiated the
conflict between the old ways of kin-based ethics expressed in parental or
elders' authority and the new desire for individual choice.

Such was the matter of marriage. One of Adesanya's clients,
Brother Bankole, was in a dilemma. Both prospective brides were in
Nigeria: the first he had originally wanted to marry, but the second
badly wanted to marry him. What should he do? He came to
Adesanya with his numbers. The prophet told him that he saw no. 1
illuminated by a strong light, "like a headlight," but in a minute it had
vanished, indicating an infatuation that would not last. No. 2 was
"fair"; she had a propensity for ill-health (which Bankole confirmed),
but was a better prospect. But Bankole continued with both relationships and remained uncertain:

#17   Prayer to choose one of the two girlfriends as a wife . . .
      Vision that the second girl as revealed before would be fair as his
      wife . . .

                                                      (30 January 1970)

Nine months later, the brother was still undecided:

#18   Prayer for protection, studies, wife to be chosen, and travelling to
      Nigeria on holidays next year.
           Vision that no. 2 girl as chosen earlier is the best for him. His
      going home was also revealed before saying it, that he should be
      careful so that there will be no difficulty for his coming back
      again. Bringing the wife is not going to be easy.
                                                    (15 October 1970)

The next time the would-be husband asked for prayer about his
choice of partner, Adesanya refused: "God has already chosen."
    What is striking about this case is that despite all Bankole's visits,
the Spirit's repeated recommendations went unheeded. Mr. and
Mrs. Olawo submitted a list of potential flats to the prophet—and
ignored the result. "The visioner can't compel you," said Adesanya.
"You are the one who chooses. He only says what he sees." Visions
reveal contingent circumstances in order to inform rather than
replace conclusions reached by clients themselves. This becomes
clearer from several diary entries about decision-taking:

#19   Prayer for a sister who wanted to join her husband at home or
      continue with her studies here.
           Vision that God will give her mercy if she decided to go home
      and if she preferred to stay in London she would be prayerful for
      her course to be successful for she might meet difficulties.
                                                    (14 March 1970)

#20   A brother brings his cousin to ask whether to proceed further to
      do his PhD or stop at MSc as he is will[ing] to go home if the
      interview he wishes to attend was successful.
           Vision . . . for the cousin. If he goes home now the way is open
      for success but although if he stays to do the PhD the difficulty
      will arise and take longer time than he expected except if he prays
      very hard against this.
                                                    (2 March 1970)

In both cases, despite advice, options are left open; either path is fea-
sible if problems are met with prayer. Aladura divination deals with
individual choice, assisting clients to confront the possibilities in their
lives. There is no community weight behind the Spirit's suggestions,
monitoring compliance. By placing the decision on another plane, it
appears to have been taken out of the client's hands. But individual
agency is not preempted; it is, rather, reinforced by helping question-
ers think through the possible repercussions of their actions and how
they may intervene, strengthened by the power of the Holy Spirit.

Active and Passive Systems of Divination

The privileged insight into relevant but unobservable factors provided by revelation not only assists the consulter in decision-taking, but will suggest steps, whether ritual or mundane, to ensure the success of the project in question. Visions always enjoin prayer; when individuals seek out the help of prophets, it is said that they "go for prayer"—not "visioning." It is this necessity for action that distinguishes C&S divination from the plethora of contemporary methods claiming to draw back the veil between the present and the future, be it through the forecasts of palmists, card readers, or crystal gazers, or spiritualist messages "from the other side."[3] Although such predictions may help clients reflect on their problems and influence personal plans, they do not claim to offer a certain way of altering the foretold events, or of affecting future destinies. The factors that determine these are either unassailable, such as the movement of the planets, or remain inaccessible or unimagined. There can be no divinely informed intervention in their activities. Such practices could, therefore, be characterized as active and passive systems of divination: they delve into the realm of the hidden and unseen, anticipate events, but cannot empower the individual to alter what is to come.

The value of the Aladura system to its adherents rests on the conviction that the causative factors behind human experience can be influenced. As Fakeye explained (chapter 7, p. 148), to know the future is to have some purchase on events. God is the source of all power, whether exercised for good or evil: *Olọrun alagbara julọ*, "the Power above all others." Through the privileged insight he affords into the play of forces affecting human lives, his help can be enlisted to counteract negative influences and strengthen supplicants' own power of resistance and efficacy. I characterize such divination as "active" not only because individual agency is intrinsic to the system, but because the ritual generated by the consultation is thought to act back upon the relevant causative factors that have been revealed as lying behind the client's condition, and so change the course of events.

This distinction between active and passive practice is derived from the elders' own concern with the danger of confusing prophecy with fortune-telling. As with Adesanya's initial reservation about Mr. and Mrs. Olawo, they deplored members' misuse of revelation to look into the future without accepting responsibility for what lay ahead. Certainly the prophetic dimension of revelation was a major attraction to members: "If we pray," said one, "God will tell us what is going to happen." But elders complained that people came to C&S "thirsty for vision," and that prophecy, rather than prayer, was more sought after in London than in the Nigerian church. This was impossible to verify, but the problem surfaces in the visioners' diaries:

#21   Prophet Adesanya had carried out repeated prayer, mainly over the phone, for a nonmember worried about her final catering exam. Her husband had returned to Nigeria and had stopped sending her money. If she failed again she would be unable to earn enough to take herself and her two children home. Adesanya gave her ritual advice, and visions promised success. But immediately after taking the paper, she rang him to discover the result. The prophet refused to do this "unnecessary" prayer, and complained to me: "Some people have no faith. They think C&S is just like herbalist—you go there and see the future. It can't help you at all."

So Adesanya warned a couple whom he suspected of consulting another visioner, that nothing could improve without prayer:

#22   Vision that they should not go about chasing vision as they do not actually believe. But if they trust in the Lord, God will change their life for the better.

(1 September 1970)

That is, if they pray. "It's no good to have thousands of people in the church and just tell them vision which you then do nothing about," said another prophet. "These people need practical Christianity; vision is just minor."

In pursuing the C&S distinction between revelation and mere prediction, I am not proposing that divination systems should be subject to a binary typology of active and passive methods. In cultural practice there are numerous intermediate cases; for example, take the African-Caribbean practitioners who appeared to offer services similar to the C&S:

I have God given powers to help you and to satisfy you with your troubles . . . I have strange and mystic powers. Problems of love, marriage, business difficulties, quarrels, enemies or bad luck present no problem to me . . .

(From a flier: "Mrs Adam from
the 'West India Land of Jesus' ": 1972)

Mrs. Adam evidently shared a discourse of external causation with her clientele, and seemed to be offering individual consultations along Aladura lines. But she would credit her insight into unseen influences more to her own "mystic power" than to specific information entrusted by God. Therefore, although she undoubtedly offered personal ritual empowerment and a counteroffensive to enemies,

these would not be based on action specifically prescribed by the ultimate source and commander of all unseen influence.

She is also, of course, operating outside an institutional structure,
whether in terms of a religious organization, or as a shared body of
belief and ritual practice. Even where these are in place, they do not
necessarily predicate active divining: not all African systems illuminate
unseen causative influences in order to affect them. But the model of a
continuum, with the majority of indigenous African oracles falling
toward one end and most contemporary fortune-telling located at the
other, could be a useful tool in comparative studies of divining. An
exploration of the material and epistemological factors that correlate
with an active system are beyond the scope of this study, but as I have
suggested earlier (chapter 1), a significant feature is the degree of elab-
oration in the discourse of spiritual power. Without returning to
Horton's distinction, following Popper, between closed and open cosmologies (1967), I would suggest that the more embracing the notions
of unseen energies, the more active the divination system will be.

The key variable, therefore, in comparative studies of divination systems is the circular, dynamic connection between seen and hidden
worlds, not the technology of the oracle itself. When Zeitlyn (1987: 41)
argues that "emotive" types of divination, such as spirit possession, cannot be subject to the same kind of analysis as mechanical manipulations,
he is overprivileging the *means* of divination as the critical factor in dis-
tinguishing different oracular systems. There may be more in common
between procedure based on inspiration and one that examines con-
figurations of objects, than between two systems that employ similar
techniques. Reading the arrangement of leaves disturbed by Zeitlyn's
spiders in Mambila divination (1987, 1990) may be more in line with
spirit possession than with interpreting the pattern of tea leaves.

The indigenous Yoruba oracle, Ifa, relies on the throw of nuts;[4]
Aladura revelation on possession by the Holy Spirit. Despite these differing forms, they are both active systems of divination: armed with
forewarning of the activity of enemies, both unseen and human, ritual
action could be taken to frustrate their plans, and secure a favorable
outcome to envisaged projects. As with Aladura prophecy, this action
is integral to the system, hence the saying, "One does not consult Ifa
without prescribing a sacrifice" (Awolalu 1979: 126).[5]

The term Ifa refers both to the oracle and to the *orisa* Orunmila, the
chief messenger of Olorun, who controls the fall of the nuts. As with
Christian prophecy, therefore, the ultimate source both of the message
and of the power activated by subsequent prayer is the Supreme Being,
the legitimator of all spiritual energy, who can influence all other unseen

forces. Coming from the same Yoruba cultural frame, details of the C&S inspirational system, as we will see, echo Ifa divination. But it is the similarity in the communication between divinity, diviner and client, and in the notions of spiritual power that energize the system, that grounds the continuity between the two.

The Structure of an Active System of Divination

The interaction between deity, diviner, and supplicant can be seen as one of structured dialogue, in which four stages can be identified. In Aladura practice, the timing and location of these stages vary; they may be elided in time and place, or be chronologically or geographically distinct. But analytically they may be distinguished as follows:

(A) *Address.* Through the visioner's prayer, clients place their situation before God.
(B) *Response.* The Spirit replies through revelation, which the visioner interprets for the clients.
(C) *Action.* Ritual and/or secular action is undertaken in the light of the message.
(D) *Outcome.* Ensuing events are interpreted as an outcome of the action taken.

This interaction may be represented diagrammatically as follows:

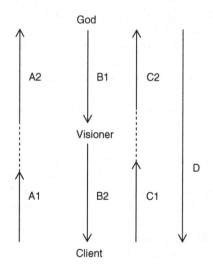

The case of the Olawo's son's illness (#10) illustrates the process. Having understood the problem (A1), the prophet puts the situation before God (A2). By means of this prayer, he gains access to the supreme source of power, who has the ability to control whatever force initially provoked the child's disorder. The responding vision from the Spirit (B1) is interpreted (B2) to confirm that the illness need no longer cause concern, given the specified ritual (C1) and (C2)—water imbued with power—that will induce the Spirit to strengthen the child and protect against future attacks. The prophet told me that the child did indeed recover, as was predicted (D).

Here we see the structure of divination praxis displayed in this recorded event: the address to the Spirit, the response through vision, the action of blessing water, and the apparently satisfactory outcome. The same fourfold structure is reflected in the operation of the Yoruba oracle Ifa. Whether the client has divulged the problem or has remained silent (A1), the *babalawo* addresses the divinity through prayer, and casts the sacred palm nuts or a divining chain (*opele*) (A2). According to their fall, determined by Orunmila, the diviner makes marks in the wood dust of the divining tray (B1). The resulting configuration refers him to one of the 256 *odu*, the named sections in a vast body of poetry in his memorized repertoire. He then recites a number of associated verses (*ese*), each one telling the tale of a mythic client's consultation—the problem, the oracle's advice, and the positive or negative outcome depending on the performance of the specified sacrifice. One of these will be taken as a relevant precedent for the present case (B2). The *babalawo*, adept in medicine as well as sacrifice, may assist with the required ritual for the client (C1) and (C2) who will then hope to see a beneficial result (D).

The ritual equivalence of the *babalawo* and the prophet is not, therefore, only due to the similarity of the services they offer; the underlying structure of their divinatory practice is identical.

The Dialogue of Revelatory Divination in the C&S

(A) Address

Adesanya's clients often had to wait for prayer when they visited his house, sitting on a sagging sofa, watching television, or chatting to his wife. When their turn came, they would enter a small front room where the prophet prayed, standing in front of his altar, a small cloth-covered table holding a sevenfold candlestick and crucifix, besides the bottles of water and olive oil to be consecrated for clients. Adesanya would be wearing his prayer gown, with a Bible, and sometimes his staff, in his hands.

His clients would kneel down before him. The proceedings would be prefaced with prayer to Almighty God. This might last a few minutes, or take half an hour, but would follow the formula of forgiveness, consecration, thanksgiving, and request for spiritual power before seeking assistance on the client's predicament. Just as the *babalawo* prefaces divination by addressing Olọrun, Orunmila, and other key divinities, so the visioner opens up the channels of communication with the Spirit.

With regular clients, such as the Olawos, Adesanya would be aware of their circumstances. New clients would sometimes be specific as to their concerns. Others, visioners told me, would launch into a confused account of their troubles, in which case the visioner would cover the salient points in the introductory prayer, hoping that the Spirit would clarify the situation through revelation. But often people do not mention all the circumstances they have in mind (#10, 18) or remain silent on the reason for their visit, requesting "general prayer." As a *babalawo*'s consulter might whisper a request to Orunmila, leaving the diviner in ignorance, so C&S clients may put their problem directly before God.[6] Members explain that they have faith in the visioner; so it is "unnecessary to go into details": "These visioners, they can pierce through into your mind." Adesanya was more cynical, claiming that members are "testing" the visioners, and "trying their power"—a notion familiar from Ifa[7]—but accepted that the ability to locate a predicament or project without foreknowledge increases the client's faith (e.g., #18). "So many of our people are still Thomasses.[8] If they have told you what they have come for, they think you are just repeating it when the message comes." Hence Adesanya's reluctance to pray for Mr. and Mrs. Olawo when they first came to a service. Just as *babalawo* claimed to perceive clients' problems through their medicines,[9] visioners had no doubt that the Spirit would guide them toward relevance, through vision, voice, *imọ* ("intuition"), or the direction of the prayer itself. Prayer, Adesanya told me, often "comes automatically, as you are directed by the Spirit. You begin to pray for one thing, but are taken onto something completely different," thus indicating problematic areas in the client's life. This dialogue between diviner and deity continues throughout the process.

Whether they are aware of the client's main preoccupation or not, visioners say that they rarely question those that arrive for prayer. Their rationale is a reluctance to inhibit an individual's own communication with the Spirit by emphasizing the visioner's role as intermediary. This shift away from the presenting problem toward the person is consistent with the holistic approach that characterises solicited prayer. It is the whole person, not just their problem, that is presented to

God: "I bring Sister Oduntan before you . . ." What C&S divinatory practice in effect attempts is a reintegration of fragmented experience. The nature of members' lives is one of compartmentalized activities in an alien environment. The impact of this disjuncture on some clients is one of disorientation. They are even unsure how to describe their trouble; they feel suffused with anxiety, engulfed by physical malaise. When the C&S prays for a brother to "make him whole," this has an existential dimension, which revelatory divination addresses.

(B) Response

Through the medium of vision, voice, or intuitive understanding (*imọ*), the visioner receives the response of the Spirit to the introductory prayer, and relays the messages to the client. As the Spirit has been confronted with the total package of a life, these may range over a number of topics (as in #11):

> #23 A sister who went regularly to the prophet, came for "general prayer of protection and development of Holy Spirit and conse-cration of water." Among the visions that were revealed was instruction "To help her husband in prayer from slumberness which can debar her own progress spiritually. Should be careful of reply[ing to] the letters received from his parents to avoid misun-derstanding and hatred. Over his exam there will always be mercy [so he] should not think that if he met disappointment that his prayer is still [not] going to be answered. The sister should be prayerful and should not worry for not having practical experi-ence enough on her qualifications in this country because God is going to provide the right job at home. Secondly, the only prayer is not to do clerical or office job for a while, as trading will open to her in the near future but should not cheat people so that the blessing of God may be upon her. [She should pray for] protec-tion and healing for their child who is now sick, etc."
>
> (16 December 1970)

From this array of revelations, as with the range of Ifa's *ẹsẹ*, the client will select the message most relevant to what he or she has in mind, while taking on board the additional information. This may bring other worries to the surface: although the Olawos asked for prayer for their child, what they actually wanted to talk about was a problem with relatives (#10). When a message does hit on a current preoccu-pation of the client, this increases confidence both in the person of the prophet and in the process of revelation and prayer. As with those consulting Ifa, it is at this point that clients may open up. "Vision really brings introduction to the matter," said a prophet. "There are things people don't like to say but, when the vision comes, they just voice it out." When Adesanya told a Mrs. Ofalabi that he saw her weeping, the sister broke down and told him of her husband's infidelities. If supplicants are ready to discuss their situation, the visioner, as the *babalawo*, can both listen to their problems and offer advice.

Consistent with the holistic approach to clients, the visioner will make use of the Spirit's messages as to other aspects of their situation, attempting to unearth hidden relevancies and connections. Sparked by revelation, an experienced visioner—with a cooperative client—may try to unravel the complexities of personal lives, and uncover the root of the trouble, the basic factors behind their malaise. As Adesanya said: "What they may take as the main trouble may not be. There may be other things leading to it. They want you to touch what's in their mind at that particular time, but you must help them understand the cause of their problem as well as find solution. That's what it's all about." An entry in another prophet's diary recorded the visit of a man having difficulty selling a house co-owned with his wife. The visioner saw the couple standing under a stormy sky, buffeted by the wind. He therefore prayed for reconciliation between the couple rather than for a supply of potential buyers. Another of his cases concerned a woman complaining of violent headaches. During the prayer, Psalm 51 was revealed to him. Being a psalm used for forgiveness of sins, the prophet divined that the cause of her suffering lay with her own behavior. When I asked one sister why she went to visioners for prayer, she replied: "Because they tell you what is the matter."

What rather than *whom*. Despite the personalization of experience that the concept of enemies entails, visioners claim that they do not set out to name aggressors, even if their identity is revealed in vision:

> #24   After prayer for healing and the consecration of a house, "Vision revealed: He should be aware of [a] friend who actual[ly] wants his destruction. Beware of bad ways that are not spiritual as this is one of the ways they can catch him. He must be sure of what he takes [i.e. eats], not to mistakenly take poison . . ."
>
> (25 February 1970)

The friend, even if depicted in vision, remains anonymous. The most Adesanya would do is to confirm or question a client's conviction that an uncle's curse or a former colleague's jealousy lies behind their problems (#13). Often these suspects are at home, in Nigeria, and therefore outside the London context. Systems have to suit their clients: many of the African divination systems that aim to identify the malefactor operate in fairly bounded communities, characterized by fixed, multiplex relationships, which must be mended if awry. If witchcraft is a metaphor for acrimony, it must be withdrawn or neutralized if community life is to continue.[10] In contrast, the C&S is a voluntary organization in the diaspora; personal accusations would disrupt the fragile cohesion of the congregation, alienating potential adherents and undermining the problem-solving project of the church.

Fixating on an identified enemy would also absolve the client from personal responsibility. The brother in #24 was cautioned about his own bad ways as well as deceitful friends. The diary entries show that although external causation is much in evidence, visions (as *odu*)[11] also counsel care in personal relationships (#10, 11, 23) and encourage clients to look to their own behavior (#11, 18, 23, 34). In diagnosing problems, revelation often throws the weight of God behind basic home truths:

> #25   A brother comes to enquire "whether to take his exam or withdraw as he is not able to prepare very well for the exam.
> Vision: . . . The remaining time before his exam if he worked harder and don't postpone the exam, God will so mercy as if he leaves it till another time he will still have the same complain[t]."
>
> (5 August 1970)

Others anxious about their exams received even shorter shrift:

> #26   . . . running about [with his social life] will not help him in his studies.
>
> (5 December 1970)

> #27   He should not slumber in his studies as his house is now crowned with laziness.
>
> (7 January 1970)

Neither does an awareness of impinging malice prevent a perceptive prophet from interpreting what he is shown as worldly distractions and inner insecurities:

> #28   A sister came to "conquer hard luck and unsettlement of mind . . .
> Vision that the problem with her boyfriend will be settled between now and 1971 . . . God will help her to settle and pass

her exam and should have a bottle of water consecrated for
this. Furthermore she should not neglect her studies for any
unnecessary things as her understanding is minimised since she
came over [to England] for her studies which should not be so.
She should avoid comparing or judging herself with others so
that she would not misjudge herself and then lose her talent . . ."

(13 October 1970)

The possibility of victimization is suggested by the prescription of
Holy Water, which would boost her spiritual immunity to attack.
But it is the sister's relationship to her own self-image and her lack of
mental focus that is primary. In the reading of experience offered by
revelation, quotidian, physical, mental, and spiritual factors
interpenetrate and fuse; just as human malice can activate spiritual
attack, so invisible assault has physiological and psychological
repercussions (#11):

#29   A sister and two friends came for general prayer.
     "Vision: . . . The first lady should consecrate water for safe
delivery and for God to conquer bad dreams for her. There is a
spider web on her face, which reveals the obstruction to her spiritual eye and thinking caused by the enemy to make her doubt
her ways and brought [i.e., bring] her failure . . ."

(4 November 1970)

In this vision, the ensnaring thread of malignity is detected as critical
to the sister's situation, but the relationship between the enemy
and her failure is not a simple one; it is the clouding of her own
spiritual strength that distorts her perceptions and undermines her
self-confidence.

Reflected in the prophet's entries such as these is the lack of rigid
demarcation between the material and intangible world already discussed. Even in the discourse of external causation the secular and the
sacred inform each other in indeterminate degrees. Take the ubiquitous notion of poison, for example: the balance between actual toxic-
ity and spiritual venom remains ambiguous (#8, 11, 24). Even in the
central concept of enemy, and the use of means to imbue malignity
with performative power, notions of supra-human activity and effect
elide with human expressions of aggression and vulnerability.
Adesanya's clientele consisted of men and women whose perceptions
of the relative weight of spiritual and mundane were neither fixed nor
uniform. This is reflected in the discourse of revelation and prophecy
and in Aladura ritual practice.

(C) Action

As with visions revealed in services, most messages include the explicit injunction to "pray fervently" or "continue in prayer." Belief in the efficacy of prayer, fortified by ritual acts and symbols, is the basis for this third stage in the divinatory dialogue, in which client and visioner render back to God what he requests. Consistent with the understanding of religion adopted here as a discourse of spiritual power, ritual in the C&S context has to do with the purposeful accessing and use of this energy. The emphasis in Aladura ritual praxis is on empowering action. The C&S occasionally talk of "believing in" their theological tenets, but they more often speak of "doing doctrine," employing elements of their ritual repertoire to strengthen prayer, to "produce results."

Special Apostle Abidoye's *Ready Made Prayer Book* opens with the "Purpose of Prayer" (n.d. [1993]: 1):

(1)   The most important lesson we can learn is how to pray . . .

(4)   Prayer surmounts or removes all obstacles, overcomes every resisting force, and gains its ends in the face of invisible hindrance. We must persist in prayer, much of our prayer fail for lack of persistence which is of the essence of true praying . . .

(5)   Lord Jesus Christ, teach me to know thy name, and to be able to use it as a key to open all problems and solve them. Let your name be my Access card[12] for all purposes to heal the sick, to loosen those tied . . .

(7)   . . . prayer is the masterkey . . . what are you waiting for pray and pray until you get what you need.

Whilst revelation is the means by which the Spirit contacts visioners and clients, their voice in this dialogue is expressed through prayer. But Aladura prayers are more than communication; they are offered to God in the expectation of exchange; they aim to initiate the mobilization of power on the supplicant's behalf, giving them the edge in social and spiritual relationships, and so to "get what you need." The traditional sacrifice of living beings and inanimate objects, with the gift of their power to God to rectify relationships, is now supplanted by powerful words and ritual paraphernalia; the hens, goats, and money of former times have been replaced by prayer and rituals aimed directly at the Holy Spirit.[13]

For members, the ritual items that lend extra charge to prayer will be familiar from Sunday services: the use of candles, the consecration of water and olive oil, together with the recitation of psalms and Bible verses. The Spirit may specify, in revelation, the particular ritual required:

#30   Prayer for one of the evangelist's children who will be travelling for protection . . .

[Vision] for the Evangelist to pray with Psalm 121 over water for the child to bath so that the Glory of God will be cover him wherever he may be going.

(9 June 1970)

If unspecified, the visioner will prescribe appropriate action from his own experience. This may be effected immediately but, if the required ritual stretches over several days, the client (as those sacrificing to Ifa) may return for help:

#31   Revelation by one of the prophets [in a Sunday service] that the husband needs seven days prayer with fasting to pass his forthcoming exam and wish to hold the prayer with me with seven candles.

(20–26 October 1969)

Fasting is thought to intensify the impact of prayer (#13, 31, 35). As a means of purification and as a personal sacrifice to God, this discipline empowers both the person and the supplication. "K&S, and Christianity, depends on power of efficacy of prayer, and this depends on fasting." (Apostle Abidoye). During Lent, all members are expected to refrain from eating or drinking from midnight until 6 p.m.; those preparing to lead the service or participate in a "special prayer" for a particular purpose, may fast for three days. As Prophet Fakeye commented: "You can't have real power without fasting unless you are using juju."

For juju can be used for self-advancement as well as attack. As we have seen, spiritual power is in itself morally neutral. Whether of indigenous medicine or Christian prayer, it may be used for whatever purpose. Despite his injunctions against "victory prayer," Abidoye's booklet is unequivocal: "To Conquer Devil and Devilish People" goes beyond protection from malignity to annihilation:

. . . All the power of Satan is stopped and cut from flowing to you my enemies. You shall die. Your source of power is cut off hence you have no more power. Michael the Ark Angel is here to strike you until you die . . .

(*Ready Made Prayer Book* 14)

In the discourse of prayer, therefore, a reliance on faith and God's will coexists with an instrumental confidence in the ability to access power for whatever end. The Spirit's power may be solicited—or hijacked.

These contrasting understandings of ritual action are reflected in the range of meaning in ritual symbols. When looking at C&S metaphors for spiritual power (chapter 5), we saw that there is an ambivalence surrounding the performative effect of those powerful objects and utterances represented in vision. So with ritual items at work, accompanying prayer. The range of interpretation, expressed both by ritual practice, and through direct exegesis by members, encompassed both traditional concepts of instrumentality as well as more mainstream Christian meanings where the symbol represents a communion with God. In C&S practice, the more manipulative approach on the part of a prayerist was signaled by a proliferation of ritual requirements attached to his or her advice on prayer. There also would be an emphasis on what these symbols could *do*, reminiscent of the ingredients in the *babalawo*'s pharmacopoeia, when he acted as *oniṣegun* (adept in indigenous medicine) as well as *oloriṣa* (priest). Conversely, a simpler ritual practice conceptualized the sacred item as an aide-memoire, both to the supplicant and the divinity, but not as indispensable for effective prayer.

These instrumental and expressive meanings are exemplified in one of the many C&S pamphlets brought to London from Nigeria:

CLAPPING, KICKING, DANCING AND SINGING
These are the mode of happiness by which this order was founded and when a member practices these in a service it is believed that he becomes endowed with different kinds of blessings appartaining to each action which are revealed by devine injunction as follows:

Clapping symbolises—Victory (and it brings it)
Kicking symbolises—spiritual power (and it brings it)
Singing and dancing symbolise—Joy and Happiness (and they bring them)

HALLELUYAH HOSANNAH & SALVATION (IYE) CHORUS
These are songs of praises to God and they as well [are] very powerful and do immediate wonders . . . [such as the collapse of the walls of Jericho].

(ESO C&S, *The Order* n.d.: 49)[14]

Aladura booklets use the terms symbolize, signify, and represent in ritual exegesis, but as this text indicates, the metaphorical and performative interpretations of the connection of ritual to power, even if distinguished, are not mutually exclusive. The various items are said to be significant for concentration and inspiration; they are also said to "do the job."

A range of meanings, therefore, may be expressed at different times by one individual, whether a prophet or his client. Take, for example, the use of psalms; elders would tell you that psalms are recited for inspiration and concentration in prayer. But when they warned of the potential perils of psalms, the verses were clearly conceived of as possessing a potency of their own. Pastor Tomori was always eloquent on the potential danger inherent in ritual symbols:

> If you read the psalms, you will know that not all of them are for praises, so you have to be very careful . . . When you give some psalm to some people, you are likely to be the first target upon which such a fellow will try it. And when you are working with somebody, and you don't expect it, he will shoot you.

Private instructions to members to write a psalm or Biblical verse on a piece of paper to be put under the pillow or carried around treats the text as talisman, as does the conviction that an appropriate psalm can produce immediate effect, promulgated by the Nigerian Aladura booklets brought to England:

> Psalm 70. To meet a case in court. Read this nakedly at night with Ps.35 and 40 three times each on a cup of water and drink.
>
> (Sonde n.d.: 12)

An area where the tension between metaphorical and performative power was often made explicit was in the use of Holy Names of God (*oruko Olorun; oruko mimo*). These forms of verbal symbols are often linked with the citation of particular psalms (#5, 11, 30), or portions of the Bible. Psalm 46, recommended to Mr. and Mrs. Olawo (#11) is advocated in a C&S pamphlet thus:

> If your wife or people in your house like to fight often among themselves, read this Psalm into water and let all bathe with it. All will be at peace with one another. ADONAI, ADOJA, JEHOVAH to be pronounced when reading these psalms.
>
> (Akindele n.d.: 13)

In his own booklet Abidoye recommends: "To Heal a Sick Person . . . Read Ezekiel 37, Verses 1 to 13. JAH-SAPADONAKAH-ILIKIM" (n.d. [1971]: 22). However, in his subsequent *Ready Made Prayer Book* (n.d. [1993]), he argues that the name of Jesus Christ is a sufficient "Access card." "Jesus is the source of the New Power brought about in the Gospel," so unlike those in the Old Testament who "needed some names to give them strength to do or perform the task before them . . . we do not have to invoke other spiritual power, nor call any Holy Name, nor seek other intermediaries." (5).

The ambivalence around Holy Names arises from the lack of consensus over their relationship to spiritual power. On the one hand they are seen as analogous to *oriki*, adulatory Yoruba praise songs to deities and mortals, where the citation of attributes implies personal knowledge of the person honored, generating obligations on the part of the praised.[15] So familiarity with his names implies relations of reciprocity with God. By reminding the Almighty of various aspects of his power, Holy Names are held to induce him to respond to the particular prayer in hand—*Alanu* (merciful one) for forgiveness of sins; *Alabo* (protector) if enemies threaten; *Jehovah Saratahubi-Sahulah* (God who revives, or enlivens) to consecrate water, and in all circumstances *Jah Jehovah* (the Almighty living God, God of Power):

> Holy Names are more or less giving praise or shall I say flattering. Because it tends to give praise to God before you . . . ask for something from him . . . Even in our own native ways of doing things, if you want to do something you call on that thing in the right way. So, if your nickname is so-so-so, they use that name and shower praise on you because they are asking for something, and I think that they have some evidence that it works.
>
> (Teacher Oluwole)

On the other hand, Holy Names are often imagined as "working" because they will automatically mobilize aspects of the Almighty's power, just as Yoruba incantations, *ofo* and *ogede*, activate the other ingredients in an *onisogn*'s recipe: "It takes you straight to where you want to go; it makes you quicker" (Aladura Brother Wilson). They are also, for some, packed with intrinsic performative power themselves: "*Ofo* are like Holy Names; they are a form of expression that will make things happen—magical words for things to change" (Prophet Korode). For visioners, the most powerful Holy Names are those revealed to them personally by God; although Adesanya might explore his repertoire when praying alone for his clients, he would not disclose such secrets in their presence. Esoteric titles are not to be divulged, for Holy Names (especially if mispronounced) are potentially dangerous wild cards in novice hands: "If you have a son, you don't give him a knife—that means you give him a weapon to kill himself" (Pastor Tomori). But even familiar Holy Names can act as effective protective charms: "JAH! JAH!! JAH!!!" read a card on an elder's dashboard.

This elision of instrumental and expressive meanings typifies all the principal ritual symbols in the C&S, employed in private consultations as in services. The use of candles in prayer is ubiquitous, and the number significant. One reminds the supplicant of the one true God;

three, of the Trinity; seven, of the seven elders around the heavenly throne.[16] But they are also said to activate these spirits of God to respond to prayer. Some go further: the positioning of a candle lit in the prescribed manner is a "powerful" technique for "victory," and in an emergency, a red candle "will work straight away." Prayer gowns and incense represent purity, a state that already has performative implications as a precondition for personal power. But they are also said to be "very useful": the smell of incense, and the armor of conse-crated white cloth automatically repels evil spirits, attracts angels, and so protects the user. Holy water (*omi mimo*), the "medicine of C&S" (#8, 10, 23, 28, 29, 30), is used in many traditions for healing physi-cal and emotional ills. The bottles, found grouped around altars both in the church and in private rooms, are imbued with power through insertion of an elder's staff, and carried back home by members for drinking or bathing. So empowered, the person may then resist antag-onistic influence. But it is also "like weapon"; water can be blessed for a specific end, its power defined by the purpose.[17] When blessed by a powerful person or in conjunction with particular ritual aids, the intrinsic strength of water can be used, according to the comparative principle of power, to force one's will over a weaker entity. Anointing with consecrated olive oil for healing reinforces the power of both people and prayer, whether through achieving a closer relationship with God or imposing effect. Any of these may be recommended to a client by a visioner, who will advise the client on ritual behavior, but is unlikely to expound on the precise dynamics of spiritual power involved. Ritual symbols are simply presented as things used as an aid to effective prayer.

The epitome of symbolic instrumentality is that curious collection of occult prescriptions contained in *The Sixth and Seventh Books of Moses, The Original Key to the Sixth and Seventh Books of Moses, The Greater Key of Solomon, The Lesser Key of Solomon*, and so on.[18] Although they include the uses of psalms and lexicons of Holy Names, the use of "books other than Bible or hymn-book" was denounced to clients and congregations alike, as I discovered on my first visit to C&S. The majority of elders condemned those who "instead of going into Bible, rely on books, which is just like running to Devil to bor-row his power." One critic explained to me that with the use of these conjurations, formulae, and paraphernalia, "the results may come even more quickly—but when they come through prayer, they are more strong and lasting"; when they are "forced," they will be "without foundation." "If you want supernatural power," said another, "just ask your God."

But there were also those who, although not openly admitting the use of "Six and Seven," were more equivocal about spiritual esoterica:

> . . . It is left to you, both sisters and brothers, to be able to find out for yourself—struggle for life's existence, struggle for the mysteries, struggle for the hidden power . . . Some people say: "Oh, God is everywhere, he can answer me at any time." But it is not like this; you have been deceived . . . Although you have pray[ed] several times, but you have found out that hardly you get reward, hardly you get five per cent of your prayer . . . which are answered, because you have not prayed in the proper place, in the proper manner, at the proper time . . . That's why we have so many rules and regulations in the spiritual field, that once you go against one, then probably your prayer will not be answered, probably it will not go up . . . You must know how to combine all these things to make it work.
>
> (Prophet Ayorinde. Sunday School 31 August 1969)

On this prophet's altar table in his living room, besides the bottles of water and olive oil brought by his clients to be blessed, there was often a pot of honey or perfume to be consecrated for a particular blessing. There might be a bottle of ink, or an egg inscribed with Holy Names for an anxious student.[19] Critics maintained that the ritual he prescribed for those who went to him for prayer included invocations and instructions well beyond standard Aladura requirements. "You cannot bring way of magician into the way of God," commented one. Ayorinde was well aware of this disparagement, quoting Biblical precedent for ritual prescriptions,[20] operations,[21] ingredients,[22] and signs [23] in his own defense: "They do not understand. If you try and explain, and do these things here, people will say you are a witchcraft."

The fact that ritual meanings were implicitly contested is not a weakness of C&S practice, but a strength. As Bell observes in her study of ritual, ". . . evidence suggests that symbols and symbolic action not only fail to communicate clear and shared understandings, but the obvious ambiguity or overdetermination of much religious symbolism may even be integral to its efficacy" (1992: 184). The evidence she cites is from studies relating symbolic practice to the reinforcement of sociocultural solidarity. But it could be argued that this lack of consensus also operates to strengthen the *ritual* significance, the epistemological efficacy, of the symbols themselves. By tolerating ambiguity around symbolic instrumentality (as with the explanatory weight placed on malign spiritual influence), C&S ritual is surrounded by an interpretive space, to be filled by individual nuance. This allows a wider constituency to trust that ritual is indeed doing what it sets out to do—to engage the power of God in human affairs.

Lack of logical internal coherence of beliefs is by no means peculiar to Aladura; it is a universal propensity in everyday human thought, which is rarely put to the test by the collation and comparison of its components. But the institutionalization of inconsistency—whether in regard to the significance of mystical attack or the instrumentality of symbols—is particularly significant for the C&S. Adesanya's clients were men and women on the move, existentially as well as physically. Among the high proportion of nonmembers seeking his assistance were established elite professionals—doctors, lawyers, politicians, and academics—looking for an effective solution to their predicaments while cautious or even skeptical of "pagan" thaumaturgy. Even those clients who were professed Aladura were inconsistent, seeking the empowerment offered by indigenous practice, whether "native" or Christian, in varying degrees, and at different times in their lives. To be successful in London, as in Nigeria, Aladura had to accommodate those whose world was alive with spirits and powers, as well as those whose universe was more secularized, that is, less quickened by unseen energy. There could be no fixed epistemological consensus, or an accepted degree of mystification of experience, for C&S clients were not only at odds with each other in interpretations of reality, but also within themselves.

(D) Outcome

The final stage in the divinatory process is the outcome of the consultation. Immediately, Adesanya's clients might have experienced a sense of reassurance, have felt relieved at the chance to unburden problems and seek advice. How far their consultations served to alleviate anxiety, and therefore truly increase chances of success, is impossible to measure. But the naming of fears, the unpacking of diffuse anxiety into its components, may well have significant impact, helping clients to feel they have a purchase on their problems and some control over their uncertain lives. The presence of the prophet, as someone who listens and understands, may be as therapeutic as what he says, but his overall message is also one of reassurance and personal support (#2, 4, 8, 10, 12, 13, 16, 22). Although tempered with realism, visions are never defeatist:

> #32 A sister comes "looking for the change of work hoping to have
> an interview on the following day."

    . . . [she] was told [in vision] not to put hope on the interview,
"but God will soon open another way for her and the job will be
good."
                                        (31 October 1969)

Another sister:

#35   . . . should not be afraid over any dream as God will always con-
      quer for her.
                                        (17 December 1970)

The encouragement contained in vision will often be reiterated by
the prophet, himself convinced in its authenticity both through his
belief that he has been guided by the Spirit, and by his experience of
the apparent impact of prayer. He also, like the *babalawo*, may seek
direct confirmation from God that the ritual sacrifice has been
accepted. But what he also has to do is to communicate these certainties to his client. Divinatory visioning does not work unless the client
sees the Spirit as moving behind every stage in the process, including
the events that succeed the session. Of the process of prayer and revelation, Prophet Aiyegbusi said: "You have to have faith if you want it
to work for you." This faith is not just a matter of morality; a C&S
client must come to interpret subsequent events as an outcome of
prayer, even when they do not appear to confirm revelation or bear
evidence of the superior power of God.

It is this process of coming to experience life in terms of certain
presuppositions that Luhrmann (1992: 312) calls "interpretive
drift"—the slow, often unacknowledged shift in a person's manner of
interpreting events as he or she becomes involved with a particular
activity. There is nothing exceptional about this: analysts' patients will
begin to interpret their lives in terms of unconscious processes forged
in childhood; Luhrmann's neophyte magicians in contemporary
Britain came to see the world as enchanted. For the C&S, it was a
process of sharpening perceptions on the operation of spiritual power
behind their daily experience, whether in increasing sensitivity to
external causation, or the possibility of protection through the power
of prayer. Given the ubiquity of notions of spiritual power in Yoruba
epistemology, the shift is not so great as for British witches, who must
internalize a hermeneutic structure that runs counter to mainstream
secularity. But C&S clients must still be convinced that what happens
to them after consultation with a visioner is connected to the ritual
action they have performed.

Although rationalizations around the effects of prayer are never presented as a body of intellectual belief, they are produced piecemeal to account for apparent failure in the system, as well as its success. For whatever the dismissal of fortune-telling, much of Aladura divination is prophetic; the test of a revelation's authenticity is that it "comes to pass." Visioners gain reputations for predictive accuracy—hence a client's tribute to Prophet Korode: ". . . when he dreams it is seldom the opposite." Zeitlyn (1990: 659) argues that "in the case of emotive divination, the truth of divinatory results is guaranteed by the possessed state of the diviner . . . [so] mistaken practice is impossible." Not so, especially when it comes to prediction. Aladura allow for error, but incorporate such explanations within the self-corroborating logic of the system itself, the "secondary elaborations of belief" made familiar by Evans-Pritchard's classic study of Zande oracles (1937: 330). In C&S prophecy, Adesanya emphasized, "all these promises are conditional"—on the actions of the visioner or the client, or on God himself.

First, the *visioner*: even when in Spirit, visioners do not relinquish agency. The simplest explanation for a visioner's inaccuracy is that he or she is lying: "saying what you have in mind rather than what God has shown you," a deception that God deplores:

> The prophets prophesy lies in my name: I sent them not, neither have I commanded them: they prophesy unto you a false vision and divination, and a thing of naught, and the deceit of their hearts.
>
> (Jer. 14:14)[24]

Aware of the weight that revelation carries, visioners are warned to speak with care. "Hearsay" visions based on gossip and "prepared" vision planned by interested parties beforehand are denounced:

> #34 . . . as the prayer was in progress, I saw three great swords, suspended, and these swords were white and very sharp. So God is warning, before God and Man, that we must be very careful what we force out of our mouths so that the church will not be devoid of blessings for everyone. And when that one passes away, before God and Man, God is warning all visioners against false vision, that we must be careful of prepared vision . . . if anyone says anything which is not true he will be punished severely. Pray that in years to come, the wrath of God will not be upon you.
>
> (Evangelist Somefun. Service 7 November 1970)

False visioners, elders maintain, are "automatically cursing themselves, and the punishment will fall back on them." This is said to range from loss of spiritual power to blindness or "sudden death" (cf. Deut. 18:20). How far visioners do fabricate their messages is hard to say. But it seems probable that the majority act in good faith, for they too depend on revelatory divination, not only as practitioners, but as consumers; amongst those visiting Adesanya were visioners who had clienteles of their own.[25] But the possibility of falsehood provides a potential explanation of predictive inaccuracy, as does the most common charge—of misinterpretation of a message. A visioner may, for example, mistake the recipient of the message: Adesanya told me that he had wrongly identified the gender of a pregnant sister's baby, but she had come for prayer accompanied by another woman, also "expecting," who therefore must have been the intended subject of the vision.

Or it may be the tense of the message that is problematic, with either the visioner or client "getting the past, present and future all muddled up." As few visions specify exact dates, "what you think is false vision may come to be true later on." When an elder sister gave birth to a girl after receiving several dreams and visions that she was to have a boy, she explained to me that the revelations clearly did not apply to this child; she would have a boy in the future. In theory such mistakes could raise doubts as to the visioner's spiritual strength, but they do not necessarily do so in practice. For although a speedily fulfilled vision is taken as proof of prophetic power, so also is the ability to predict a distant event.

But it is not only the visioner who bears the responsibility for apparent prophetic failure; it may be the *client* who is at fault. A promised blessing may be withheld, it is said, because of sin, ritual impurity, broken vows, or lack of faith. Specified action may have been neglected: water not blessed, candles not lit, fasting shirked, public thanksgiving evaded, prayers not said. Perhaps the client has failed to make the advised personal effort in work or social relationships? A further explanation for an unfulfilled promise is the client's changing situation. Predicted exam success cannot be enjoyed if the candidate falls ill, cannot study, or fails to sit the papers. So a London prophet wrote to a member in Nigeria reporting a revelation that concerned him: he was to become a doctor and lead a team at Ibadan University. Some months later the prophet received a letter: why had God made such a promise? Financial problems had forced him to abandon medicine, and change to a non-degree course in another subject. The prophet replied that it was the change of circumstance that had rendered the vision redundant; a request for further prayer should have been made directly this became apparent.

Lastly, the ultimate rationalization for unrealized predictions lies with the *will of God* himself. Members are constantly reminded that God is the best judge of members' interests, such as in the time frame of their progress. Teacher Oguntulu had been in England since 1962, so was encouraged by several visions he received in 1969 that he would return to Nigeria at the end of the year. He did indeed pass his final exams, and prepared to go home. But financial and domestic difficulties delayed his departure until 1973. He explained that the revelations had not been wrong; God had seen that "the time was not yet right." This was confirmed by subsequent visions of the dangers of unemployment, illness, accidents, and so on if he had returned. It was also revealed that he was delayed "for a purpose" and that he had "further work to do in God's field."

But most often it is the effect of prayer that is held responsible for an outcome of divinatory visioning at odds with the prediction. God, as with the threatened destruction of Nineveh (Jon. 3, 4), can be persuaded to change his mind:

#35   I bring another fact to you today, I reveal another truth to you, so saith the Lord of Hosts. When my anointed Prophet Fakeye was in your midst, it was revealed that he should pray that he would not be in this country six years before leaving. Have you forgotten the revelation? The prophet had three days consecutive fasting without telling anyone, and so the true God answered his prayer, and turned the revelation upside down. Why can't you then, ye sinners, ask in the same manner, and your bad vision will be changed to good?
(Prophet Korode. Vision. Service 18 January 1970)

So, when exams are passed after visions warned of disappointment and an ill-omened journey proves trouble-free, "this does not mean that vision is wrong; vision is just a warning." In public testimonies and private conversations, members will recount the signs they have received of answered prayer: a narrow escape after predictions of an accident; the rescue of an upturned paraffin heater after visions of a fire. For if a warning is heeded with prayer, "God will say OK, but he will send signs to show that he is not telling lies and that he has stopped it from happening." The fact that large numbers of predicted problems do not arise serves to confirm, rather than undermine, the efficacy of the divinatory system. Fulfilled visions demonstrate the power of prophecy; unpredicted outcomes prove the power of prayer. It is only when Adesanya's clients come to interpret the events that follow the prophet's visions in the light of these assumptions, that revelation can become active divination.

The Role of the Client in Revelation

The active participation of the client is therefore as crucial in individual consultations as it is in public visioning. As Peek (1991b: 3) points out, if divination is approached as a dynamic process rather than as a diagnostic product, the critical role of the divinatory congregation/client comes to the fore; Aladura revelation is no different from other African oracles where the process of creating knowledge proceeds through the dialogic relationship between consulter and diviner.[26] Both *babalawo* and visioner depend on their clients for *selection* of the message. Just as those who whisper their request to Ifa's palm nuts pick out the appropriate *ese*, so clients who have not yet confided in the visioner will indicate which message touches on their problem. In revelation, as with Ifa, the additional visions/*ese* may be accepted as extra information if deemed relevant by the client, or allowed to fade from memory if not. It is only clients or congregation members who are skeptical of an Aladura worldview who will dwell on errors and ridicule inapposite visions.

In the *amplification of meaning* of the message and *interpretation of the text* (see chapter 7, pp. 155–60), the client's willing participation is as essential as in Ifa divination. To receive advice about an "intention" (#16) can be taken as an adequate reference to a project, while given the latitude surrounding the meaning of metaphor and the generality of many of the messages, the client can read relevance into a range of forecasts. The same is true of Ifa: although the *babalawo* assists in decoding the *odu*, it is his client who will extrapolate the meaning from the mythology. As a Yoruba proverb states: "The Ifa oracle speaks in parable, but the intelligent will follow" (Adesanya 1958: 37). When Clarke (1939: 241–242) queried the relevance of a predicted death in the family to a man quizzing Ifa about sweepstake results, he was told: "On the contrary; if we interpret death as a loss, the connection is close." C&S members also speak of the allegorical or allusive nature of revelation; as all who consult diviners and fortune-tellers, they will wrest meaning from the message through their own interpretations. This may only emerge later; it is the client who, with hindsight, discovers prophetic value in messages previously received. An accident, burglary, illness, or other misfortune will have been satisfactorily predicted by a message to "pray against any obstacle in your way"; a financial windfall or an examination success by a promise of future blessing. At the time of the consultation, once the message has been selected and interpreted, and metaphorical references fleshed out with personal circumstance, the client may then provide *corroboration* of the message to the diviner. It was Mrs. Afolabi who confessed the reason for her "weeping"; the prophet had only diagnosed distress. Although not consciously articulated as guiding principles of revelation, the reinforcement of individual agency by the Spirit is reflected on the operational level: every stage in the consultative process necessitates the active involvement of the client.

The Democratization of Yoruba Divination

The parallels between the practice of Ifa and C&S divination are substantial, but, when it comes to the role of the diviner, the distinctions are marked. In contrast to Ifa, where divination was the preserve of professionals who commanded secret knowledge, Aladura appears to offer the opportunity to everyone. With rare exceptions,[27] *babalawo* were male; now women were consulted too. In terms of daily existence in Britain, there was nothing to distinguish visioners from other Yoruba students, as they combined their services with secular life. Prayer is free, so they did not depend on divination economically, but could choose "how deep" they wished to go; in London, some decided to minimize the amount of precious time they spent on prayer; others, working their way up to the grade of prophet or prophetess, sacrificed both work and study for the Spirit. Training consisted mainly in regular participation in church life and participation in prayer for others. The only formal teaching was the monthly meetings of the Visioners Band, a group attended by any spiritualist (i.e., one who falls into Spirit), or those wishing to develop their power of dreaming into visions. Here they learned to channel their possession into revelation and to control the Spirit according to C&S conventions. Entry into the Band was by recommendation of elders, irrespective of gender or age. Despite the lure of the esoteric, the official emphasis for spiritualists was on straightforward prayer and fasting. Dialogue could be established directly with the Spirit at any time, in any place, with no mediating priest or paraphernalia: "a direct telephone to God." This ritual simplicity enabled a novice spiritualist to move gradually into a specialist's role, while basic procedures were simple enough to be recommended to consulters. As a result, as Adesanya pointed out, the dependence of clients on the diviner was reduced:

> I don't want them to come to me just like *babalawo*. I don't expect them to come for vision's sake but to believe. So, when they have problem next time, they don't have to run to me, but can solve it themselves through prayer.

The greater ritual independence of C&S members compared with a *babalawo*'s clients is mirrored theologically by the direct access to Olọrun through visioning, dispensing with a mediating spirit. What Aladura represents, as a replacement of Ifa, is a democratization of power and its deployment appropriate to a new clientele. At the time of Independence and during the subsequent two decades, the hold of indigenous diviners on Nigerian society was waning. The potential devolvement of divinatory abilities to a larger proportion of the community suited a geographically dispersed immigrant body in the diaspora, as well as at home, providing both easier access to specialists and the possibility of self-help. In the practice of Aladura revelation, the indigenous principles of hierarchy, gerontocracy, and gender enshrined in Ifa were loosening, opening the opportunity to practise divination to young men and women not yet advanced in their church careers. The increasing emphasis on personal achievement in developing the power to divine reflected the mobility and individualism of men and women striving to transcend their class through educational qualifications and bureaucratic employment.

But if divinatory aptitude was so accessible, how is it that these Christian diviners in the C&S retained their authority? Since the purpose of C&S possession is oracular, the prestige of the diviner must be maintained in order to provide an effective service to clients. If their "intellectual and imaginative capital"[28]—their divinatory knowledge and spiritual power—is so democratized, what safeguards their credibility? As Evans-Pritchard noted of all departments of Zande "magic": "faith tends to lessen as ownership spreads" (1937: 183).

What was striking about C&S practice was the disjuncture between theoretical access to the Spirit and the exercise of revelation. At the beginning of 1970, there were 40 members on the Visioners Band's roll, including dreamers as well as visioners. But only those who came to meetings could speak out in church, and attendance at meetings was usually no higher than 15–20 members. The membership of the Band was also divided into four grades, only the first two of which were permitted to speak in services, as "most of the visioners are still young in spirit and they do not know the proper ways and manners of presenting visions to the public."[29] Although in what was seen as a successful service, the pews would vibrate with possession, with scores of members' shouting out in Spirit, the proportion of these who actually conveyed coherent messages to the congregation was small. Between October 1969 and May 1970, when an average congregation numbered 250–300, 12 men and 6 women spoke in church. Never more than 11 different visioners stood up in any one service;

usually less. On most occasions (such as in the services described in chapters 6 and 7), the great bulk of visions came from a handful of male visioners in senior grades. Although C&S ideology emphasizes the democracy of the Holy Spirit—"anyone can see vision"—in practice, messages reached the congregation through a tiny, predominantly male, minority. An even smaller proportion of these were sought out by others for prayer: novice visioners might relay messages to friends during private prayers at home, but in this period only two women and eight men had regular clienteles. When the sister "holding two sticks" came to decide to whom she would go for prayer, her options were few. The enduring notions of spiritual hierarchy and comparative strength of spiritual power, although cutting across secular criteria of age and status, secured the diviner's authority despite the theory of universal access to the Holy Spirit.

This page intentionally left blank

Chapter 9

Epilogue: Empowerment and Yoruba Christianity

At the beginning of the twenty-first century, does the Cherubim and Seraphim still satisfy the religious needs of the Yoruba diaspora in London? The majority of Nigerians now living in Britain are of a different generation from the worker-students of the 1960s; conditions have changed, both of the Yoruba community and in the wider society. The religious scene too has been transformed by the explosion of new churches, hostile to Aladura. How has the C&S weathered the competition? And are these burgeoning organizations as radically different as they claim to be, or are there underlying continuities in Yoruba religious practice?

Samuel Ajayi's Daughter and Her Contemporaries

Evangelist, now Senior Special Apostle, Ajayi has retired from the Ministry of Mines and Power, and, as often as failing health allows, comes to England to visit his daughter. Bose Amusan is a tall, elegant woman who is making her life in London. Born in Lagos in the 1960s, she was educated in private schools and trained as a nurse. In 1991 she followed her husband, a solicitor, to England. Although Richard was educated in Nigeria and Germany, he was born in England; so they both have secure status as citizens. Their small East London house is not affluent, but their daughters go to a local independent day school. Not for them the fostering arrangements of the earlier generation; the children are fetched from school by a Czech au pair when Bose attends her midwifery course at university. Her decision to leave Nigeria is not unusual; although one of her brothers is a Lagos lawyer, the other two, both doctors (one trained in Nigeria, the other in Poland), now work in the United States.

The Amusans are typical of their generation: young professionals or businessmen and businesswomen with higher education. Some were born and brought up in Britain; some returned to study, often after sojourns elsewhere in Europe. Some came to England for the first time in the late 1970s and 1980s, financed by family or a government rich from oil. Others followed later: as the oil boom collapsed in 1981, and economic and political life degenerated into chaos and corruption,[1] the idea of emigrating for work and study as well as the education of their own children became increasingly attractive. Some of this second generation already have their parents in London; several of the older men and women in the C&S also returned to the United Kingdom when unable to find work or secure a proper pension in Nigeria. Although here I am crudely contrasting two generations in London—Yoruba worker-students and their adult children—the distinction, of course, is not a rigid one. But it is one articulated by Yoruba themselves. Older Nigerians such as Ajayi recognize that their children are better educated, more affluent, and more widely traveled than themselves, with a more thorough command of the English language.

The Yoruba community they join is more firmly established than that of 30 years before. First, over recent decades London has witnessed an increase in the ethnic minority population at large, both through immigration and natural growth. Larger numbers and longer-established communities have given minorities a greater purchase on national life. Second, the profile of the Yoruba diaspora itself has changed. The exact number of Yoruba in London is unknown, but the 1991 census showed that between 1981 and 1991 there was a 50 percent increase in the Nigerian-born population in Britain.[2] This has been enlarged by the British-born: over a third of West Africans had been born in the United Kingdom.[3] A Nigerian magazine recently put the number of Nigerians in Britain at 1.5 million,[4] but this represents a new sense of significance in British society rather than a realistic estimate. Research based on the primary language of London school children conservatively suggests that there are at least between 51,000 and 56,000 Nigerians from the two main Nigerian ethnic groups in London (Igbo and Yoruba), 43,000–47,000 of whom are Yoruba.[5] As nearly four-fifth of the Black African population lived in Greater London in 1991,[6] total numbers of Yoruba in Britain might be around 70,000.

The Yoruba community is by no means homogenous—there are those who struggle for a livelihood as car park attendants, underground station cleaners, hospital porters, and taxi drivers—but through their income and education many Yoruba are clearly distinguished from the poorer sectors of black minorities. These more successful men and women are no longer generally perceived as part of an immigrant proletariat as were their predecessors, but as members of a black middle class, which has consolidated over recent decades. Yoruba men and women own shops, businesses, and commercial networks, and offer a gamut of professional skills and services. There are Nigerian-run Internet sites, newspapers and magazines, and a Nigerian channel on Sky TV. The size, self-sufficiency, and confidence of the Nigerian community acts as a buffer between itself and British society to a much greater extent than at the time of the first C&S congregation.

But as members of a black ethnic minority, Yorubas' experience of Britain is still problematic. Whilst contemporary Yoruba now move more easily within mainstream London life, this does not mean that they feel wholly at ease within it. Immigration legislation has been progressively tightened since the 1970s, and many Nigerians fear falling foul of it. The Macpherson Report (1999), instigated by the murder of a young African-Caribbean man, confirmed that racism is ingrained in the institutional fabric of British society, and expressions of personal prejudice, from insult to violence, are still rife. Bose has held senior nursing positions in NHS hospitals with a wide range of staff, but it is likely that she so often mentions her friendly relationship with white neighbors because, in her own experience, it is rare. Tensions between the African and African-Caribbean communities have now also been publicly acknowledged by black commentators themselves.[7] Nigerians may no longer have to suffer stereotyping as illiterate savages, but they have acquired a popular reputation for arrogance and fraud.[8] Although cushioned by their own community, they still have to contend with contemporary versions of worker-students' problems—financial, psychological, physical, and personal—and negotiate a Yoruba identity in a white society.

One difference between the mobile post-Independence generation and their parents concerns this identity and its relationship to Nigeria. Even if some never achieved it, worker-students aimed to return. Many of the younger generation who have recently arrived in England are leaving their options open, with permanent settlement as a strong possibility. This suggests two contrasting ideal types of immigration. Worker-students of the 1970s, aiming to return home after temporary residence in Britain, would seem to exemplify a circular pattern, whereby people remain identified with the place from which they came, while the contemporary diaspora could be seen to illustrate a linear model, a shifting orientation away from Nigeria toward gradual absorption by British culture. A young Yoruba professional typified his generation as "sharing a mind-set, an interest in Western culture and values."

But lived reality is more complex. Those of the earlier generation who are found in London (who stayed, came back, or travel between Africa and England) do retain strong links with Nigeria. They certainly display a greater cultural orientation toward their country of origin than do their emigrant children. But at the same time, these one-time worker-students have accommodated themselves to London and their reference points include both societies. Similarly, a simple linear model distorts the affiliations of the younger generation. Although they may, like the Amusans, have left West Africa to settle in Europe or the Americas, many still identify with their ethnic background. They may still retain strong family connections in Nigeria, reinforced by modern methods of communication and mutual visits. It remains to be seen whether their own children retain multiple identifications. The Amusans' family life includes many Yoruba referents—language, music, clothing, food, and friends—and this will influence the way their children imagine themselves. Some British-born grandchildren of the worker-student generation may feel themselves to be Yoruba, while others may only be perceived as such because of their surnames.

Simple models of the complexities of contemporary migrant experience and ethnic identity therefore will not do, but some generalizations relevant to their religious culture may be suggested. First, both former worker-students and their adult children retain a Nigerian identity to some degree, but this is considerably stronger amongst the older generation. Second, especially for younger men and women, a Yoruba identity is no longer predicated on West Africa alone. The Yoruba diaspora has spread not only across Africa, but through Eastern and Western Europe and the United States. Yoruba connections do not all lead back to Africa, but are filtered through diverse locations. Through their global networks, the present Yoruba community in London has a more transnational identity than that of their parents' generation.

Taking a historical perspective, these distinctions between the generations represent continuity as well as difference. Earlier I argued that there was a correspondence between the generation that founded Aladura in Nigeria and that which brought the C&S to Britain. Both were migrants, one from the countryside into the towns, the other from Nigeria to Britain. Both were the first generation to achieve a certain level of education, one in school to qualify for white-collar colonial service, their successors in college to prepare them for a professional life in independent Nigeria. Both were dealing with expanded ethnic horizons, first in pan-Yoruba Nigerian towns together with other national ethnic groups, and for the later generation, in cosmopolitan Britain. Both generations were bent on moving up the social scale: the first, in the first half of the twentieth century, into the petty bourgeoisie; the second, into the immediate post-Independence elite. For each, a vital component of consolidating their changing class position was their children's education, equipping first their sons, and then their daughters too, to advance even beyond their own achievements. Both generations represented the modernity of their day.

Generalizations are always undermined by individual exceptions, but these same patterns are clearly visible in the biographies of many of the following generation, those of Bose's contemporaries. As the elite of the Nigerian nation-state or migrant professionals, the arena for their own geographical mobility and experience of multi-ethnicity is now global; their education takes place not only in colleges, but also in universities, national and international. They too look into the future and consolidate their class position by attending to their children's education. The adult children of C&S worker-students now stand, as did their parents and grandparents in their time, for the modern world.

What, then, of the church? Does this new generation feel identified with Aladura? Are these younger men and women joining the C&S? Do they feel that it can help them achieve their ambitions and lead successful lives?

Aladura Revisited

Bose's childhood was steeped in the C&S. Her mother is a devout Aladura; her father became ever more senior in his church. As a young girl, she went to live with her maternal grandparents. Her grandfather had been in the Praying Band with Moses Orimolade Tunolase. Yet during this period she left the C&S. In London, Earlham Grove is within walking distance of her house, yet the only time she will go there is by invitation for a naming ceremony or a similar celebration. This is not often, for few of her acquaintances are Aladura.

The dearth of members of Bose's age was one of the things that struck me when I returned to Earlham Grove after some 20 years. The hall itself had been impressively improved; the new furniture, decorations, musical equipment, and sound system bore witness to the donations by Bands and members over time. The members' prayer gowns were resplendently elaborate. But the congregation was smaller than I remembered: by the end of the service 30 women and 17 men, including visitors in ordinary dress, sat in the body of the hall, with a further 15 or so in the choir and band. The elders on the altar platform were just that—10 men, each one well over 50, several of whom I recognized from the congregation in the 1970s.

The service progressed along familiar lines, but undelayed by spontaneous choruses or manifestation of the Holy Spirit. The elders around the altar retained complete composure throughout the service; among the congregation I saw only one sister shaking, and shouting out a message. When a female member announced the item on the program for visions, this was unheralded by prayers or hymns to summon the Spirit. Only a senior prophetess stood up to recount what she had seen, speaking in Yoruba to the congregation in general rather than any individual. A couple of messages were briefly translated: the church must pray against misunderstanding with the youth, so that they may come into the church. Looking around, I saw that the congregation was mainly composed of parents and children. There were almost no young adults present. The following sermon, given by an apostle invited from another branch of the C&S, incorporated several conventional Aladura themes: the divine purpose behind problems; the victory vouchsafed by God in the spiritual warfare of life; the threat of enemies disguised as friends. But the chief message concerned the personal cost to the faithful in following Christ, as sheep among the wolves of skeptics.[9]

The church announcements, read by the church secretary, an elder sister, preceded preparations for an impending naming ceremony. Although most of the congregation appeared to be Yoruba, an elder offered a brief explanation of the way in which the symbolism of the water, salt, and honey placed in the child's mouth, related to traditional rituals to bestow blessings for a successful life. With the naming ceremony and the celebration of the birth of the baby boy, the service became livelier, with the dancing and spontaneous choruses that I remembered from the past. When the service ended many seats were still unoccupied, but the familiar commotion among those present showed that the social significance of the church for its members was as important as ever.

This was confirmed by visits to other branches of the C&S. Some, where the halls are packed with over 100 exuberant members, including some younger men and women, reminded me of Earlham Grove in its heyday. If the original headquarters is now comparatively small, it is because several of the elders and visioners have, over the years, broken away to form their own churches.[10] The C&S Council of Churches, created in 1976 to counter the fissiparous nature of the movement, has 37 C&S branches on its books, but elders' estimates of the actual number of churches ranges from 40 to 60. While some of these are small, ephemeral groups meeting in private houses or rented rooms, others are well-established organizations that have purchased their own properties.

There is much mutual visiting between C&S branches, but this "mushrooming" remains a source of grief to some older elders. Special Apostle Abidoye, who, moving between England and Nigeria, heads the C&S Council, laments the trend in his recent booklet: "I wish people would stop proliferating new churches and join any one of the existing churches. But," he adds, "do not forget he or she would say God told her or him in vision, to start a church who am I to say no" (2001: 44). The summons of the Spirit may be the Achilles' heel of a unified church, but it has always been the main method of the spread of Aladura. Other Aladura organizations—the Church of the Lord, the Celestial Church of Christ, the Christ Apostolic Church—have similarly splintered during the last three decades.[11] But just as this process of fission disseminated the church throughout Nigeria, so the myriad branches of the C&S have become a familiar part of the burgeoning scene of black-led churches in Britain. Aladura are still alive, well, and expanding.

The C&S as an AIC—an African Indigenous/Initiated Church

Although events precipitating fission are local, the process of fragmentation has been heavily influenced by the schisms that have shaped the Nigerian church. When C&S leaders from Nigeria came together in London to celebrate the dedication of Earlham Grove in 1971, several took the opportunity to create branches of their own factions in London. So Captain Abiodun ordained Brother Olusanya, an elderly GP, as senior apostle, to form a branch of her own C&S Society, which he still heads as St. Michael C&S Church in Plumstead. The Spiritual Father E. Olu Coker[12] (of the ESO of the Morning Star Fountain of Life, Mount Zion) elevated Lady Leader Awojobi to be a prophetess of his church. Now entitled Senior Apostolic Mother, Awojobi still leads her own small congregation from her cavernous Victorian church in Crouch End, North London. Evangelist, now Baba Aladura, Somefun also converted his prayer group into a church (ESO of the C&S Thursday Prayerist Church, UK), affiliated with his old Nigerian branch, ESO Hotonu Street. In 1978 the church purchased a large old chapel off South London's Old Kent Road. Although there are a number of small unaffiliated branches, the majority of London churches are associated with one of four main C&S groupings: the ESO of C&S, the ESO of the Morning Star, the C&S Society, and the C&S Movement.

Despite its early connection with the Northern Movement, Earlham Grove has resisted Nigerian affiliation. It was Special Apostle Fakeye, the founder of a highly successful C&S Movement branch in Lagos, full of returnees, who prompted the formation of the Holy Order of C&S Church Movement (Surulere) No. 1 in London. This was in 1986, and Fakeye, like many of the elders of the 1970s, is often to be found visiting London. But the cycle of splits began again with the division of the church in 1993. There are now some seven Movement churches, led largely by men and women who were origi- nally part of Earlham Grove, whereas others have moved out of the Movement umbrella and remain unaffiliated. The influence of the Movement is also felt in Britain through the administrative organiza- tion that characterized its origins. Over the last three decades, the spreading Yoruba diaspora has carried the C&S with it, and the Movement has established international structures to coordinate its branches in the United States, Germany, Denmark, Italy, the Irish Republic, and Britain and liaise with the World Conference Headquarters in Nigeria.[13] These arrangements are loose and allow for autonomy, but provide a forum for mutual visits by elders. The other groupings are also developing similar international networks.

Structurally, then, despite the national and international spread of the C&S, and independent local developments, the church retains strong links with Western Nigeria, constantly reinforced by the movement of elders across the Atlantic. The persisting Yoruba charac- ter of the church is reflected in the ethnic composition of C&S con- gregations in London. In theory, the C&S is as keen as it ever was to recruit from all nationalities; in practice, despite a minority of African-Caribbean and non-Nigerian African members, congregations remain largely Yoruba. The greatest involvement of an African-Caribbean contingent has been in Birmingham, where there are now several C&S congregations. But the bitterest schism, in 1979, was largely along ethnic lines, between the Yoruba leader and a rival grouping led by a Jamaican pastor.[14]

Efforts have been made at ethnic ecumenicalism on the Aladura side: Father Olu Abiola (son of the late C&S Baba Alakoso,

A.A. Abiola, in Nigeria) not only founded a small multi-ethnic church named Aladura International, but also a Council of African and African-Caribbean Churches, which includes the C&S Council of Churches and other West African churches in London, besides other black religious groups. In turn, this body is a member of other organizations that include mainstream denominations as well as black-led churches. This gives the C&S legitimacy in the orthodox church scene, as does their membership of both the British and the World Council of churches. But organizational affiliations have not diluted Aladura identity as an AIC, an African Initiated Church. When in the autumn of 2001 the C&S School of Theology (an evening Bible class) wanted to consolidate itself by joining forces with others, it was the leader of a Ghanaian spiritual church, the Musama Disco Christo, to whom it first turned.

At the same time, C&S elders recognize that what is seen as the African aspects of their practice deters the younger generation; they know that their AIC is seen as too indigenous. The C&S preacher's lament and the prophetess's vision on the lack of youth speak to a deep concern: there are still small children in their little white prayer gowns, but few members from their late teens to their thirties. The failure to recruit the younger generation in significant numbers surfaces painfully in C&S meetings:

> Why are our people running away? . . . Where are we going wrong? We want to retain our children . . . if there is anything which will not allow our children to stay with us, we must look at it. We've got to move forward.
>
> (Special Apostle Abidoye, C&S Council of Churches 20 January 2001)

One concern here is the use of language. Members of the younger generation are not all fluent in Yoruba, and whilst the worship in some churches is muted by the attempt to squeeze English translations into diluted Yoruba rhythms, other branches still combine both languages for prayer, and clap and dance to Yoruba hymns. Another is the sophistication of the music. Active choristers and a good range of musical equipment attract younger members, but the bands of many branches are rudimentary. But the impediments to participation that are most often raised are issues around ritual and the body: the removal of shoes, covering of women's hair, and, most especially, the menstrual taboo. Many members feel that the first two injunctions may be seen as oldfashioned and inconvenient by younger people, and the third is criticized also by non-Aladura as standing in the way of women's worship. Even more problematic is the prayer gown. Members say that not only is this considered unstylish by their sons and daughters, but that it firmly ties the C&S into the white garment (*alaso funfun*) tradition. This has come to stand for a wide range of practice in Nigeria, including the more disreputable end of the ritual market and, as elders say, the popular demand for "magic." They speak of the "cultural baggage" that must be left behind to attract the youth, yet the prayer gown debate seems particularly irresolvable: "We are in a cleft stick here, because this is our identity."

So young Yoruba men and women may no longer see their own identity reflected in that of the C&S. One sister's reply to Abidoye went straight to the heart of the matter:

> Where are our children? Look at us! What have we got for our children to engage them? They are with Born-Again!

For over the last decade, another version of Christianity has been carried to Britain by the Nigerian diaspora. Younger Yoruba are turning to Pentecostal Born-Again churches, which, despite their Nigerian associations, are seen as both structurally and theologically more liberated from an African past than their Aladura precursors.

The Challenge of the Born-Agains

Bose and her family do not have far to go for church. On Sundays at noon, only a couple of miles away from Earlham Grove, a congregation of some 2,000 mass into a large auditorium converted from a modern warehouse on a Hackney industrial estate. In the huge car park stands a fleet of minibuses for connection to public transport. The building itself, with its spacious carpeted reception area, bookshop, meeting rooms, and offices, has the feel of a corporate headquarters, but inside the equally well-appointed hall the volume of clapping and singing celebrates the religious nature of the business. The sweeping semicircles of chairs, with raked rows around the edges, continue to fill throughout the afternoon with African, African-Caribbean, Asian, and Black British worshippers, plus a handful of white participants. Estimates made by members reckon that 50–60 percent of those "fellowshipping" will be Nigerian, the majority of Yoruba origin. But 40 national flags hang on either side of the large central stage, announcing the cosmopolitan membership of the Kingsway International Christian Centre (KICC).

Most of the smartly dressed congregation, a few in Nigerian clothes, are in their twenties to forties, but there are also older men and women. Younger children are in the crèche, but there are scores of style-conscious adolescents in the hall. However street-wise the teenagers may be, they have no embarrassment in immersing themselves in worship, shouting in spontaneous collective prayer and obviously familiar with the songs and choruses that fill most of the five hours of service. The rear of the stage is taken up with the choir, a carefully lit line of red-robed men and women, each with a microphone; on their right are drums, electric guitars, a piano, and two keyboards. One of these is played by a woman, as is the saxophone. Gentle musical meanderings often provide backing for prayer. A female vocalist leads sections of the singing. The style of the music is neither gospel nor hymnic, but of well-rehearsed contemporary popular lyrics with a strong rhythm. The well-known words of the choruses are sung over and over again:

#1 I went to the enemy's camp, and I took back what he stole from me; Satan is under my feet.
#2 We are champions . . .
#3 Victory is mine, Victory is mine, Victory today is mine.
    I told Satan, "Get thee behind me"; Victory today is mine.
#4 I just want to praise you
    Lift my hands and say I love you.
    You are everything to me
    And I exalt your holy name . . .
#5 By thy great power . . .
    Nothing is too difficult for thee . . .
    Nothing, nothing, absolutely nothing,
    Nothing is too difficult for thee.

The premise on which this confidence is based is a total commitment to Jesus. Each person must repent of the past and hand over the present and future to the Spirit—so being Born-Again. The guide on this journey is the Bible, illuminated through constant prayer. The duty of a "new Christian" is to "minister," to bring others to "know God."

There are no ritual symbols anywhere in the auditorium: no cross, no candles, no vestments or vessels; nothing to identify a religious or cultural tradition. Attention is rather focused on the pastors, both men and women, who lead the singing and prayer. All speak in English, but when dealing with matters of God, rather than promoting meetings, prayer groups, conferences, and KICC "products,"

American inflections often dilute their Nigerian accents, betraying U.S. evangelical influence. The key figure in the KICC is the founder and leading pastor, Matthew Ashimolowo, adept in evangelistic techniques. When preaching, he uses the whole stage to dramatize his hour-long address. With carefully timed sections to vary the mood, he drives home his principal theme with exhortation, allegory, role playing, humor, autobiographical anecdote, Biblical text, and prayer, underlined by constant repetition both by himself and the congregation, of the key words in his central message.

The theme of one typical sermon (24 September 2000) was the Holy Spirit, "the road to unlimited success" in marriage, career, business, financial matters, ownership of land and property, and personal relationships. The preacher reiterated the promise of chorus #5: "You will not fail." With the anointment of the Holy Spirit, obstacles will fall away, opponents crumble, for "the power behind me is more than the power against me." Through the Holy Spirit in your life, he promises "a change is going to come!," and you will be omnipotent.

The whole service, although characterized by fervent participation, is a masterly professional performance by the pastors, who dictate the pace and emotional pitch of the proceedings from the center. Whereas order in a C&S service is maintained through elders' collaboration, Born-Again services focus more narrowly on the few, who hold the hall's attention through their personal charisma. The dramatic focus on the individual in the conduct of the worship is reflected in the central message of the proceedings. Despite the dominance of leaders and the size of the gathering, the appeal of the pastor to each man and woman present is personal; *their* lives, *their* experience, *their* souls are what matters. Close-circuit television completes the message: hand-held and mounted cameras, some operated by women, send alternating images of the pastor and members of the congregation onto large monitors suspended around the auditorium.

At the same time, as the cameras pan round the hall, this virtual relationship between floor and stage serves to weld the hundreds present into a collectivity. First-timers are welcomed into the community and given a glossily produced pack (*Raising Champions, Fulfilling Dreams*), with introductory fliers, CD, an elaborate brochure (*Winning Ways*), booklet (. . . *for New Christians*), and an invitation to a pastor's lunch. Mailings will follow: "Welcome! . . . after today you are part of the family . . ." The *Sunday Bulletin*, printed each week, advertises the groups and services on offer to the "brothers" and "sisters": the Caring Heart Fellowship; Bible Study; meetings for youth, women, and singles; counseling on personal, financial, legal, and career issues. Those who "give themselves to Christ" can live much of their lives within church circles.

The KICC is one of the biggest Born-Again organizations in Britain, with branches in Wimbledon and Luton. Ashimolowo, a Yoruba who attended a Nigerian Bible school before coming to London, is one of the leading lights of the Born-Again movement, which is attracting younger Yoruba both in England and Nigeria. The rise of the Born-Agains in Nigeria is part of the explosion of charismatic Christianity throughout developing societies.[15] Latin America, Asia, Africa—estimates of the profusion of churches and their evangelical success cannot keep pace with the rate of growth. Cox (1996: xv) reckons that one in four Christians is now Pentecostal. The original impetus came from mission activity, largely from the United States. Although the Pentecostal movement later divided along racial lines, its black origins at the beginning of the twentieth century bequeathed a common legacy of dramatic worship and a thaumaturgical spiritualism. This then developed according to the local context; in every case, the movement has been molded by particular socioeconomic, historical, and cultural specificities while retaining a distinct character. With its clear imperatives of the importance of the Holy Spirit and right living, its lack of prescriptive dogma apart from the moment of conversion, and its reliance on personal familiarity with the Bible, the message has become indigenized whilst swelling a global crusade.

Because of the resulting heterogeneity and the myriad churches and groups that form and reform, Pentecostal genealogies are never simple. In Nigeria, sections of the Aladura movement were influenced by Pentecostal missionaries in the 1920s; more Pentecostal and charismatic inspiration from Britain and the United States stimulated revivals during the 1960s, and the formation of new, largely indigenous, religious groups during the 1970s and 1980s.[16] Against a backdrop of national materialism and corruption, the resulting churches were often of the "Holiness" type, preaching a millenarian message, while condemning the world with its consumerism and lax morality. Younger men and women became involved, and the social character of the congregations then began to change as Pentecostalism spread amongst students and their teachers, as well as poor laborers and clerks. Education had expanded dramatically after Independence: in 1960 there were less than 1,500 students in two universities; by 1973, six universities were catering for over 23,000 students.[17] This campus Christianity became focused on small groups, which might stay to worship inside a mainstream denomination or expand into a church or "ministry" around a leader.

In the last quarter of the century, the Holiness character of many of the groups began to shift toward the "Faith" movement filtering in from the United States.[18] This embraces aspects of worldly ambition, conducive to its largely educated clientele, and is concerned with the Spirit as a present reality more than a future hope. Rebirth in the Spirit is the avenue to riches material as well as spiritual, hence the popular name for this Pentecostal tendency: the "Prosperity" gospel. A typical representative is a Yoruba Civil Engineering graduate from Ilorin, David Oyedepo, founder of the Living Faith church. His Biblical exegesis includes the message that social and sexual strictures are still important, but do not inhibit personal success, which will proceed from a right relationship with God. Religious ministries in Nigeria now range from rigid morality to unashamed materialism, with many combinations of the two. The stress on prosperity departs from the tenets of traditional Pentecostalism, but Born-Again convictions are also influencing mainstream denominations.

The evangelical zeal of Born-Again Pentecostalism has resulted in the establishment of several churches, such as the KICC, by Nigerian pastors in London.[19] These are not identical; they vary in size, and espouse a range of Pentecostal messages. For those attending Born-Again services, the style of worship and confidence in their own "rebirth" appear to be almost more important than any particular theological thrust. Dominant pastors also have their distinctive style. So Pastor Albert Odulele of the Yoruba-founded Glory Bible Church preaches extravagant Prosperity, whereas his co-pastor Jonathan Oloyede has a more Holiness perspective. Drawing on their experience with university Born-Again groups in Nigeria, these two men have gradually developed a small "house church" into a church of several hundred strong. They meet in a Newham warehouse, which was purchased in 1997 for £1 million and is now converted into "The Glory House—Where Miracles Happen." Despite this range of emphasis, and the overt Holiness doctrine of organizations such as the Deeper Life Bible Church, the Faith perspective predominates in these new Nigerian churches. Ashimolowo's preaching on the Holy Spirit as the "road to success" is typical of the KICC. A similar message is conveyed in E.A. Adeboye's Redeemed Christian Church of God (RCCG), established in London in 1994, whose network of large London congregations is part of a global membership estimated at over a million.[20]

Yoruba also attend others of these new Pentecostal churches with a different ethnic provenance. When the Amusans first came to London, they frequented The Kensington Temple (KT), a large multinational venue in Notting Hill, popular with West Africans. With Colombian roots and a British pastor, the KT has a strongly evangelical bent, for only those born again can be counted as Christians at all. The Universal Church of the Kingdom of God (UCKG), originating among the urban masses in Brazil, espouses a florid demonology, countered by a strong "deliverance" and healing practice involving exorcism as well as prayer.[21] The UCKG attracts hundreds of the less prosperous amongst London's ethnic minorities, but still includes a large proportion of Nigerians in its two main churches at Finsbury Park in north London, and Peckham in the south of the city.

It is by churches such as these that the C&S feel threatened. Those of Nigerian origin have a wider ethnic appeal than the C&S, but roughly two-third of Glory House and RCCG members are still Nigerian, some 80 percent of whom are Yoruba. Yet these churches do not identify themselves as "African." Through their insistence on their multi-ethnic membership, they place themselves in an international context, part of a global movement that encompasses many individual churches. Yoruba Born-Again luminaries constantly visit London, but on the Faith circuit it is the network rather than the particular church that is significant: Oyedepo (Living Faith) came to the Glory House in 2000; Adeboye (RCGC) preached at the KICC conference the following year. This activity is aimed at working the international scene rather than exerting guidance or influence from Nigerian headquarters, as in the case of C&S secessions. There is a branch of KICC in Lagos—but this was set up by Ashimolowo from London. This British-based pastor has also "ministered" in Canada, Portugal, the Netherlands, and France, and is planning churches for Ghana and the United States. The C&S, as we have seen, is also spreading overseas, but the international scope of Born-Again operations is on a different scale. On Sundays no chance is lost to refer to leaders' recent or impending travels: Europe, Africa, Asia, Latin America, and, above all, the United States. These churches are as their adherents: younger, more affluent, more mobile, less African centered than their predecessors. Their horizons, their structure, and the style they project are of global organizations, rather than Yoruba churches in the diaspora.

Born-Again Pentecostalism as a Global Religion

When Born-Again Yoruba Christians characterize their religion as "more modern" than the AICs, this orientation toward a global market is part of what they mean. Globalization, as the emergence of transnational institutions and movements, is nothing new—the world religions are a case in point, whilst economically, the international circulation of capital has permeated the postwar world. But as Appadurai (1996) and others have argued, the interrelation of mass migration and the explosion of contemporary methods of communication have transformed both the speed and geographical range of interconnections. Few societies now stand outside the reach of the global economy and its cultural tentacles. Imaginatively, too, the earth becomes a smaller space: "Globalization as a concept refers both to the compression of the world and the intensification of consciousness of the world as a whole" (Robertson 1992: 8).

A paradox of globalization is that this increasing universality is accompanied by greater particularity as individuals search for ethnic, cultural, or religious categories with which they can identify. Yoruba Born-Again Christians can participate equally in the Pentecostal "family" wherever they might be. The "modernity" of the original Aladura movement lay partly in its pan-Yoruba character; but the national flags that decorate the Glory House and KICC celebrate a wider world. As with the C&S, the preponderance of Western Nigerians in the congregation provides a social network and a sense of national identity, but the projected image of Pentecostal global connections is contemporary, flexible, and cosmopolitan, situating the individual in an imagined international community.

The cultural style of these two movements, Aladura and Born-Again, epitomizes the developments in what signifies the modern: whereas colonialism in West Africa promoted European models of modernity, it is the influence of North American capitalism that suffuses the global age. In the process of globalization, economic, political, and technological strength makes the United States a major player; interdenominational North American Pentecostal culture has deeply influenced the Born-Agains, not only in their origins, but in their contemporary style. Training of would-be pastors in U.S. Bible colleges, visits by U.S. evangelists bearing their books and videos maintain the American connection; the professional use of electronic media links Born-Agains into a transnational network of Pentecostal Christianity. The new Nigerian-led churches often have national genesis and express indigenous concerns but, unlike the C&S tradition, the presenting style that they appropriate is American.

The Prosperity gospel itself can be seen as part of this package. As we saw, concern with personal problems and ritual pragmatism lay at the heart of Aladura's "practical Christianity." But the promises of self-advancement in the Born-Again churches where Yoruba congregate are more assertively displayed. The KT's evangelical thrust is backed by the pastor's videos on *The Leader Within You* and *The Ladder of Success*. But it is Yoruba-initiated churches such as the Glory Bible Church and KICC that subscribe most wholeheartedly to a Prosperity gospel. Ashimolowo's four books on *101 Answers to Money Problems* (2000a) offer advice on "how to experience God's prosperity" through "investment, saving smartly and the purpose of real estate." Albert Odulele, preaching at the Glory House (24 January 2001), not only held out the prospect of owning land and mansions, but also of medical practices, companies with 1,000 employees, and even conglomerates. This is not the reward of some grim-faced Puritan ethic. Prosperity includes pleasure: even if goals are not yet achieved, "enjoy yourself on the way to where you are going" (KICC 23 November 2000).

If the reality of most Born-Again members' lives is less exciting than the scenarios projected by preachers, all can wander in the virtual world created by electronics.[22] The cameras and sound systems evident in the KICC service are indicative of the significant role played by modern technology in the movement as a whole. Personal religious practice involves the private consumption of expensive "products" generated by the interdenominational Born-Again network. Ashimolowo's sales brochures are typical: tapes of pastors preaching, videos of conferences, CDs of evangelical exhortation are circulated and shared among members. The oral tradition in African Christianity is translated into text. Here again, as with the social composition and pan-ethnicity of congregations, Born-Again cultural production represents a progression rather than a radical break from earlier Aladura practice, a style suited to its time. The proliferation of C&S pamphlets catered for the literacy of congregations becoming familiar with the printed word.[23] But the uneven typeface of locally printed Nigerian Aladura booklets compared with the glossy professionalism of KICC brochures provides a graphic illustration of the generation gap between the two modernities. The organizational and marketing strategies of Born-Again churches help this postcolonial generation construct an appropriate image of self that is both modern and mobile.

Looking through the illustrations in these brochures and on the covers of audio and video tapes, it is immediately apparent that they are directed equally to women and men. Many KICC female members, who make up some two-third of the congregation, visualize the plans that God has for them in much the same way as men: successful careers, financial security. Of the 12 "testimonies" printed in leaflets for a KICC Women's Conference, half concerned the financial outcome of successful prayer. This three-day meeting in November 2000, attended by some 1,500 women (plus a few hundred men), was entitled *Winning Women*. The aim was to "equip women with the necessary knowledge, gifts and talents to enable them to reach their potential in Christ, and . . . help women to break free from personal, societal and environmental limitations" (Conference leaflet 23 November 2000)—contemporary female demands in a Christian context.

At the same time, the message was also about women's responsibility for the home. One of the speakers, Mrs. Faith Oyedepo[24] (who had "just flown in from Nigeria"), reminded women that they were "the crown of God's creation." But she also emphasized that "your husband is your head," to be respected and obeyed, and outlined the domestic duties of a wife. As with the C&S, the teachings on gender are rooted in meaningful moralities drawn from Yoruba culture, which assumes the superiority of husbands and fathers. There is a difference, however. The Born-Again message complements patriarchy with the reciprocal duties of men. Whereas worker-student couples often collaborated domestically out of necessity whilst in London, their now-adult children say that this temporary egalitarianism evaporated in Nigeria. When Bose's father visits, he never enters the kitchen, any more than he does at home. But her husband Richard cooks and changes nappies. Whatever the actual shortcomings in other households, Born-Again preaching overtly emphasizes cooperative gender relations (absent in earlier C&S teaching) and underlines male responsibility. The message to the male participants in *Winning Women* was that a husband must afford his wife both consideration and support.

This is a significant advance for a woman, for, if she can persuade her partner to become Born-Again, he must abandon the culturally sanctioned male propensity toward infidelity and domestic indifference.

> I came to this church a broken-hearted, discouraged woman, ready to say goodbye to my marriage. But . . . To God alone be praise; my husband gave his life to Christ, God miraculously gave him a new job and our marriage is fully restored.
>
> (Testimony from KICC leaflet 23 November 2000)

Premarital sex is frowned upon, but single men, not only women, should observe sexual restraint, and therefore not harass younger women in the church. Within marriage, although abortion is condemned, contraception is endorsed as part of responsible parenting, which should be undertaken by both partners. These sexual and moral strictures provide guidance though the moral confusion of modernity, especially for geographically dislocated adolescents and young adults. They also enable women to negotiate a sense of self-worth when they are respected by others in the church community, including their own husbands.[25]

As Adesanya's diary showed, C&S prophets prayed constantly for marital harmony, and censured irresponsible male behavior. But in Yoruba Pentecostal discourse, it is the model of the nuclear family that is consistently promoted. Preaching to men and women whose own kin may be scattered around the globe, pastors present stable family relationships as the foundation for the careers of both wives and husbands, and the well-being of their children. Whereas older members of the C&S may have left other wives at home, or married a second on return, Born-Again Christianity does not contemplate polygyny. The mutual emotional and material support expected of Born-Again partners is another move away from the collective life of village or compound; the segregated social spheres of men and women are merging in companionate marriages.

The concept of gender relationships that informs Born-Again preaching is reflected in church organization. As we have seen, the C&S of the 1970s was already developing a higher profile for women, and this process has continued. Awojobi is not the only female church leader, and, as the Earlham Grove service evidenced, women now can assume more senior administrative roles. But at a C&S Movement conference in October 2000 a woman spoke out strongly about the subordinate role of sisters. In the KICC, women are prominent at every level, not only as office workers, musicians, singers, camera operatives, and stewardesses, but also in religious structures. Their role differs between churches, but in some, women—free from Aladura menstrual taboos—are found preaching as pastors and deacons. Although it is still usually the husband who is the head of the church, the pastor's wife (as Yemesi Ashimolowo and Faith Oyedepo) is often a highly influential figure in her own right. Both in church and secular "families," the patterns of authority reflect the changing mores and class position of socially mobile Yoruba, as did Aladura in its day.

It is not only in respect of gender relationships that Born-Again organizational structures appeal more to a younger generation of Yoruba than Aladura. The greater democracy that C&S represented over indigenous ritual practice has now been taken a step further. The powerful paid pastors, with their employed team, managing both worship and administration, contrasts with the collective voluntarism of Aladura. As in C&S, individuals can find roles in numerous hierarchical church associations. But there is no status ranking amongst the congregation itself. All members, new or old, male or female, are equivalent, reflecting their equality before the Spirit. Large congregations with no graded membership combine anonymity (and therefore personal privacy) with welcome into a community. The prototype is corporate, rather than the hierarchical style adopted in an earlier age by the C&S. To translate this into a domestic model: in both movements members are "brothers and sisters in Christ," but the C&S pattern is more akin to the Yoruba family with its ranking of gender and seniority. In the Born-Again collectivity, although patriarchy persists, individuals have a more autonomous role, as in contemporary family units in Europe and the United States.

Globalization and the Individual

What emerges from all aspects of the Born-Again message, morality, and structure is the emphasis on the *person*, appropriate for the individualism of a global age. The weakening of old collectivities is part of a process already well under way. Aladura itself was a response to an awareness of widening horizons, loosening the links between personal identity, locality, and social group. Throughout the last century, the penetration of market relations and the rights and duties of Nigerian citizenship progressively removed expanding areas of Yoruba life from norms of social reciprocities. Old moral allegiances were diluted by private ownership, legal autonomy, political enfranchisement, and—perhaps most importantly—monetized exchange. What Matory (1994: 47) calls "this new capitalist negotiation of the self" challenged previously abiding principles of seniority and gender. Even before the Born-Again generation, the socioeconomic and political hierarchies of a developing Nigeria, and then, for worker-students, of a Western industrialized metropolis, demanded a new resilience. This was not only because of a shift from ascribed to achieved status, but through the associated changing sense of self. An identity largely constructed by communal ties, whether of family, neighborhood, religious group, or economic interest, had to give way to, or at least incorporate, a conscious individualism on the part of those faced with carving out their own niche in new surroundings. Ideologically, the worker-student generation encountered the endorsement of this individualism through their education, in the notion of selfhood and free will embedded in the Christian and European tradition[26]—an increasing focus on the individual already reflected in the C&S through its democratization of structure and spiritual discourse.

In contrasting an individuated identity with an indigenous Yoruba conception of the person, important caveats must be made. Individual choice and aspiration were inserted in traditional hierarchical structures;[27] historical events and economic interests had resulted, according to Akintunde (1974: 105), in a society long characterized by "virulent competition." Nevertheless, Lawuyi (1991), another Yoruba scholar, argues that in pre-Independence Yorubaland real self-actualization was constrained by social pressures to conform. Although Yoruba sociopolitical culture undoubtedly facilitated the insidious increase in individualism, this did not, until recently, add up to the form of self-preoccupation familiar in the West.[28]

For the succeeding generation, those who now people the Yoruba diaspora, the pressures toward individualism have gathered momentum. Advanced capitalism and American and European popular culture has elevated the pursuit of self-interest to unprecedented heights; consumerism, whether the individual acquisition of material goods or of techniques of self-enhancement, is used to stake out personal identity. Born-Again publicity provides a Christian synthesis of this individualization, and the cultural complexities of what constitutes a contemporary Yoruba in London. For example, among the advertisements for African-run businesses and services in a KICC conference brochure (19–25 August 2001), a photo of a young woman with brown skin, but long straight hair, promotes *Sahara Single Bible Hair Products*. Even physical attraction, presented in an African-European package, is espoused by religion. The conference slogan on the cover of the booklet—"Take Charge of Your Family, Finances, Future by Faith"—provides a moral context for the dreams of individual progress, whilst also offering the means to realize them. The focus of Born-Again practice is individual agency, both in its goals and its epistemology. At the same time, its certainties offer a personal anchor in postmodern cultures awash with conflicting ways of being.

The Spirit and the Self

These individuating pressures of modernity, the increasing centrality of the self, infuse the "power" relationship between worshippers and their God. The whole person—psychological, physical, emotional, as well as spiritual—is the celebrated subject of Born-Again practice. We have already considered these elements in the construction of the C&S subject; we have dwelt on the importance of religious experience, and noted the sensory nature of Aladura worship. But these aspects of individual response are taken further in the new Yoruba Pentecostalism. The conscious appeal to the physical dimension of individuality is a Born-Again trait, which is foreign to C&S practice.

The aural attractions of worship are more insistent than in C&S: continuous music, both participatory and performed, whether as a highlight or in the background, is paramount in the emotional involvement of participants. This is reinforced by preaching styles, such as that of Ashimolowo, which also make direct use of tactile sensation. Those present are constantly instructed to touch their neighbors, to repeat exhortatory phrases to each other, physically connecting the congregation. Metonyms of the senses to refer to the person, the metaphor of the body for the soul, constantly appear. The subtitle to Ashimolowo's book *Breaking Barriers* (2000b) is . . . *Changing Your Speech, Sight, Hearing and Thought from Negative to Positive.* Words spoken, heard, or imagined are credited with performative power, not, as with Aladura, because they are imbued with spiritual efficacy, but because they express inner desires that Jesus can help realize. These dreams and visions are not coded messages from God, but the expression of personal goals. Sermons in all Born-Again churches emphasize seeing, voicing, hearing, walking—metaphors that externalize ambitions and stimulate the imagination: "If you see it, you can have it" (Albert Oyedepo 29 October 2000). Physical faculties are invested with power because they banish negativity and reinforce the will to succeed.

A person's efficacy is therefore associated with his or her own imagination. As in C&S thinking, individuals have their own power, and the quickening of this power by the Spirit sounds very much like Aladura explanations we have encountered. So Oyedepo writes:

> The gift of power will not come until you stir it up. There's a power deposit inside you, but it takes prayer to cause it to be released.
>
> (1996: 59)

But in Born-Again discourse, it is the individual's own *psychological* resources that are central here. C&S Prophet Adesanya, too, knows the significance of his clients' confidence and self-image, but the C&S understand innate power primarily in terms of spiritual energy bestowed by God. By contrast, Born-Again teaching draws on North American secular self-help literature, filled with "personal power" and "positive thought" to help realize "true potential." A central text in this U.S. genre, Anthony Robbin's *Unlimited Power* (1986: 29), asserts:

> We can all unleash the magic within us. We must simply learn how to turn on and use our minds and bodies in the most powerful and advantageous ways.

The influence of this "can-do" philosophy couched in popular psychological terms abounds in Born-Again discourse. Explaining the ambitious vision for his life that a newly "called" man may conceive, Ashimolowo explains that "this is because God is using what is called incentive motivation; that is using the positive to challenge a person" (2000b: 77).

There are other faculties of the mind and emotion that constitute the Born-Again agent. One such sensation that is summoned is strong affection, in the personal attachment between the worshipper and deity. The former has been individualized; so now must be the latter, in order for the emotional connection between the two to be realized. The C&S God is not completely impersonal or diffuse; he is credited with some character and with a capacity for personal communication. But his human aspect is of much less importance than his spiritual. As I argued earlier, when manifested through possession, the focus is not on the particular relationship between the host and spirit, but on the pragmatic utilization of superior spiritual vitality and discernment.

In Born-Again epistemology, this power itself has now been personalized, and invested with character. God, in all his three manifestations, is a person as well as a power. This includes the Spirit, who, KICC Pastor Ashimolowo emphasized, "is not an 'it'—he's a personality" (Sermon 24 September 2000). "The Holy Spirit is a real personality with intellect emotion and will," reads the back cover of one of his books: *101 Truths about Your Best Friend the HOLY SPIRIT* (1997). "He is the executive of the Godhead therefore, he knows and can reveal the mind of God to those who seek his friendship. What are friends for?" he continues. "Friendship with the Holy Spirit means communion, direction, intimacy and depth in the things of God . . . Ask your friend to walk with you, talk to you and take you into the depth of the Father's will . . . This is one friend who will pray, sing, speak and help needy people through you. Truly, he is your greatest and best friend." This intimacy goes beyond friendship into more emotional realms:

> I'm in love, sweet Holy Spirit, I'm in love.
>
> (KICC Chorus)

Albert Oyedepo (29 October 2000) told his congregation that "God is waiting for a passionate love affair with you."

This is not the stuff of traditional Pentecostalism, or of other evangelical denominations. In these traditions, it is the Son rather than the Spirit who is personalized; it is Jesus who is the "best friend." Jesus is not absent from Born-Again practice: the attractive African American guest artiste who inspired KICC women at their conference was not singing gospel, but popular romantic ballads where the love object was Jesus Christ; with some songs, such as chorus #4, worshippers might be addressing Jesus or the Spirit. But in this form of Yoruba Pentecostalism, the whole Trinity has been reimagined, recast in a contemporary idiom of individuality. The model here is that of the popular Western discourse of "relationships." It also has resonances of the Yoruba tradition of patron–client alliances, and the contemporary culture of negotiating Nigerian life (whether in Nigeria or England) through establishing personal connections.[29]

The central metaphor for divine–human communication is rebirth in the Spirit. The concept of being born again is inspired not by Aladura, but by Holiness and orthodox Pentecostalism, where the fastening of the bond with God, the central event of spiritual rebirth, is marked by the convert's glossolalia. The sign of God's favor in these traditions is possession by the Spirit. But to be born again in the new Faith churches rests on individual choice. The defining moment of spiritual rebirth is often highly emotionally charged, but is not necessarily an altered state of consciousness, or a spontaneous outburst of tongues. However much it appears to the person involved as an imperative dictated by the Spirit, to be born again is a deliberate, self-motivated decision to commit one's whole life to Christ. One of the characteristics of Aladura spiritual power noted above was the ability to transform. In Born-Again theology, this transformation is effected not by involuntary possession, but by the conscious choice to become a new person in Christ.

Although the dominant strain of Born-Again Pentecostalism has to do with subordinating individual will, and asking God to "take absolute control over our lives," the agency of the sentient individual in this relationship is more prominent than in Aladura. When analyzing communication with the Spirit in the C&S, we saw the spiritualist retaining a controlled, intelligent role despite the intensity of the trance experience, but for Born-Agains, no particular significance is attached to altered states. To be "anointed" is to feel the "Holy Presence and Power of God" (Ashimolowo 1999: 7), but not to be possessed. The closed eyes, uplifted arms, and radiant expression of

Born-Again worshippers appear to express deep feeling, but one that is conscious and personally managed. Their physical journey is only occasionally the oblivion and post-possession ecstasy of C&S Spirit. The force of choruses and intensity of prayer produces some shaking and trembling in KICC worship, but trance is not privileged; it is seen as a personal spiritual gift to which no central importance is attached. Born-Agains do not speak of falling into Spirit, sinking into the waiting well of unseen energy. They are filled with the Spirit, their own conscious minds and own quickened bodies aware of the Divine presence. The body is still the site for the Spirit, but in the Born-Again message to "open your heart to the Spirit," the heart is a metonymical reference to "life," to be dedicated to God. It is not an invitation for physical possession. In decentering involuntary possession from their practice, Born-Agains are further democratizing access to spiritual power, continuing a process begun by Aladura. Now a privileged association with God depends only on a personal decision.

Empowerment, Purity, and Ritual

As Ashimolowo promised, through becoming this new person, the power of God is made accessible:

> The Holy Spirit is the road to unlimited success. When the Holy Ghost rests on you, you cannot fail . . . he makes you omnipotent, to go where you have never been before . . .
>
> (Pastor Ashimolowo. Sermon 24 September 2000)

When looking at the C&S concept of power, we saw that blessing was equated with the Holy Spirit. To be blessed is a word much used in Born-Again circles: members wish each other a blessed day. As in Aladura, the term connotes success and security, and a general efficacy in life. This is achieved only through the power of God:

> You have made us more than a conqueror! Without you and your spirit, we are losers!
>
> (KICC Prayer)

Despite all disavowals of continuity with other Yoruba Christianities, Born-Again churches are still based on the ubiquitous Yoruba search for spiritual power. Hackett vividly demonstrates its significance to Nigerian preachers and their followers, arguing that the intensity of the discourse of empowerment has increased in recent times (1993: 397–402). In London, the new Nigerian churches, as much as the C&S, are dealing in the currency of unseen energy. Armed with the superior power of God, both Aladura and Pentecostal Christians believe they can forge an individual course down both familiar and unfamiliar roads, and pursue self-interest in novel settings. By surrendering themselves to a greater force, those selves emerge revitalized, to face quotidian existence. Both see the Spirit as the ultimate source of worldly success.

What, then, is the difference between the two approaches to empowerment? The line between Aladura and Born-Again understanding of the Spirit is neither thick nor straight. Apart from the experience of rebirth, many Born-Again statements, especially those published in Nigeria, seem almost identical to Aladura:

> Those born of the Spirit of God belong to the extraordinary class of people. They call divine forces into operation like the wind . . . you only see the resultant effects but you don't see, and cannot tell, how the operation was carried out . . . This supernatural life is for everyone who is born of the Spirit of God.
>
> (Oyedepo 1993: 26)

For the C&S, augmentation of personal power is achieved by tapping the spiritual vitality of God through prayer and fasting. Born-Again praxis is also based on prayer and the performance of any Christ-centered activity: attending services, reading the Bible, purchasing and listening to recordings of preaching all result in blessing. The point of Bible-study illuminated by preaching is to deepen an understanding of the will of God, and prepare oneself to accept it. The focus here is the *relationship* between the individual and the Spirit; what matters is the person's particular interconnection with the spiritual aspect of God. It is this that elicits the exercise of his power; no external, mediating, ritual aids are necessary, hence the absence of religious symbols in Born-Again practice.

The relationship with the Spirit is also considered significant in the C&S; the use of Holy Names indicates a familiarity with God in order to invoke a response. But Holy Names, as we saw, are also often thought to have performative power in themselves. Whether seen as instrumental or merely expressive, the use of ritual paraphernalia and symbolic action—reciting psalms, lighting candles, wearing white gowns—is integral to the summons of spiritual power. Pentecostal RCCG sermons on purity refer to "strict personal morals . . . [as] the central evidence of a new life in Christ" (Hunt and Lightly 2001: 17).

But for the C&S, injunctions relating to women's dress and menstruation, the removal of shoes, and the wearing of white gowns have to do with the necessity for *ritual* purity, without which the *Ẹmi Mimọ* cannot freely operate.

This notion of the performative power of ritual, its instrumental as well as expressive purpose is, as argued earlier, integral to indigenous Yoruba religious practice. Aladura Christianity has abandoned indigenous deities, but despite all disclaimers of the past, understands the operation of spiritual energies as their worshippers once did. When Bose invited me to her house for the naming ceremony of her new baby, she explained that there would be "none of those *orişa* things of the Cherubim and Seraphim." The symbols of sweetness and savor are abandoned not just because they feature in the traditional ceremony, but because they are part of an epistemology of spiritual power seen by Born-Again critics as connected with a pagan tradition. The C&S are self-reflexive on this point. Abidoye's heartfelt concern in the C&S council meeting prompted one brother to exclaim: "Look at us! We are too syncretist, too Sango-like. This is what is driving them away!"

As we saw, Sango is particularly associated with the trance-inspired power of prophecy—a central sticking point between Pentecostal and Aladura. Toward the end of 2000, the C&S approached Premier Christian Radio, an Evangelical Pentecostal program, to run an advertisement for the theological school. The station refused, claiming that it had "received complaints" about the church:

> One Lady who complained suggested that "they [the C&S] invoke curses on any of their members who stop going to their churches" and that "they see all sorts of evil visions . . . they mix pagan and Bible worship together." She said that she knew former members who have had "very nasty experiences."

The letter went on to claim that the C&S did not qualify for membership of the African and Caribbean Evangelical Alliance, who have

> raised concerns about the use of visions and revelations. In addition, many African Churches would not consider the Cherubim and Seraphim churches to be orthodox in the Christian faith and would not consider them a church.
>
> (Letter to C&S Council of Churches 21 October 2000)

Earlham Grove is not the only C&S branch to have heavily reduced the role of visioning in Sunday services. In Movement No. 1 any vision must be reported to the elders, who will take the necessary action. This is not because vision divination is no longer important. Elders still say that "vision is the foundation of the C&S"; most branches have some form of Watchnight service or vigil, and the private consultation of visioners continues unabated. But the C&S is attempting to present a public face that is less associated with African practice around the revelatory aspect of spiritual power.

Secularization and the Spirit

In the eyes of many Born-Agains and mainstream Yoruba Christians who are critical of Aladura, it is the C&S's association with this aspect of Yoruba ritual that renders the church suspect. The power of Aladura visioning is not necessarily denied: Bose recalls a C&S message that one brother would become a lawyer, another a doctor, and that she herself would marry a lawyer, come to England, and practise as a nurse. But she is nevertheless highly critical of what she sees as the potentially manipulative practice of prophecy, and relies on her own personal relationship with God.

It is this Spirit, according to Ashimolowo, who can "access you to the mind of God." "What is hidden to others will be obvious to you." Listening carefully to the Holy Spirit will "show you the secret of the enemy," give you "supernatural knowledge," making you a superior person with authority over others (Sermon 24 September 2000). With prayer, "God will tell you the strategies to use against your enemy" (Mrs. Ashimolowo 23 November 2000). This sounds like the essence, if not the language, of Aladura. But there are significant differences. What is shown is not in vision, but arriving at realization through "spiritual understanding": "knowing things not by normal way of knowing, but just knowing . . ." (Ashimolowo. Sermon 24 September 2000). Communication with the Spirit is therefore direct and personal; the mediating role of visioner has gone. Pastors do convey prophecies to their congregations, and may take dreams as significant, but revelation to individuals comes largely through their own prayer and Bible reading, not through the pronouncements of others.

What then of the preoccupation with enemies, the evildoers whose plans are exposed through Aladura divination? Is the decentering of possession and prophecy in Born-Again practice a sign that the concept of external causation has lost its purchase? Not entirely: worshippers are constantly reassured that with the help of the Spirit, they will overcome others who would set "obstacles in your path." Despite the Born-Again emphasis on individual responsibility ("Whatever happens in your life begins in your heart" (Mrs. Oyedepo 24 November

2000)), the enemy, as Ashimolowo and KICC choruses (such as #1)
reiterate, is still very much around:

> Maybe the enemy is telling you that you are a nobody, that you haven't
> got the right car, the right husband . . . It's time to take a stand and let the
> enemy know that God has great plans for you . . . Your time is coming . . .
> (Mrs. Yemesi Ashimolowo. KICC Women's Conference 23 November
> 2000)

It was already apparent with the Aladura of an earlier generation
that this adversary might be human, acting with mundane malice,
rather than an unseen spiritual foe. But there is here an added self-
reflexive dimension that was not prominent in C&S. "Satan" in
choruses (#1) and (#3) is an ambiguous force. In the same way as the
Spirit, he has been highly personalized, brought to the fore as a malef-
icent being in direct confrontation with the Christian. At the same
time, in Born-Again exegesis, he also stands for psychological negativity, both as a lack of self-confidence, and debilitating criticism by others.
These interpretations now seem more prominent than the C&S image
of Satan and his "branches," those invisible malign beings under his
control. Born-Again literature in Nigeria—as the national media[30]—is
much more explicit about evil entities, and they do still lurk in
London, where various infirmities of mind and body are rebuked as
spirits, as is "the spirit of witchcraft" (KICC, recorded phone message
11 January 2001). But in churches such as the KICC and the Glory
House, this spiritual anthropomorphism may be more a metaphor
than a reference to particular forces that populate the indigenous
cosmology. Born-Agains in Britain are less concerned with the specificities of witchcraft and evil spirits than was the generation that
sought refuge in Aladura.[31] The revelatory practice whereby the
schemes of evildoers are uncovered therefore becomes less relevant:
the divinatory project, as its agent the visioner, has been displaced.

But this is also because the cosmological foundation of the system has
been undermined. Enemies may be purely human agents because the
world in which British Born-Again Christians move is increasingly secular.
Secularization is often defined in social Durkheimian terms, as either the
decline in the power of religious institutions, or the decrease in the number
of their adherents.[32] Global technological and scientific progress does not
determine this process as inexorably as once was envisaged;[33] the worldwide
growth of Pentecostalism is a case in point. But, as the C&S preacher who
warned of skepticism was well aware, these Eurocentric definitions certainly
apply to contemporary Britain.[34] However, another possible understanding

of the term, more relevant to the argument here, is neither institutional nor behavioral, but epistemological. Following the definition of *religion* as praxis relating to spiritual power, *secularization* would signify the disenchantment of the universe, when the world is no longer articulated by unseen powers. Many Britons profess faith in nonhuman forces,[35] and hold to superstitions, fragments of a worldview that no longer maintain a logic. But the prevailing public discourse is relentlessly secular.

Members of the diaspora do not remain unaffected by this cosmological secularization of the societies in which they have settled. The coherent logic of the spiritual universe, the interconnections between specific unseen dramatis personae, the characteristics of spiritual power, and the principles by which it is thought to operate—all no longer have the purchase on the collective imagination they once did. The very term that Born-Again Christians most often use to refer to unseen energy: *supernatural* power, betrays these differences with Aladura. The C&S more often spoke simply of spiritual power, for power was part of the *natural* order. For Born-Agains in London, powers abound, but do not necessarily imbue every animate and inanimate object as *ase* and *agbara* were once believed to do. Born-Again communication with the Spirit, therefore, cannot be analyzed as an action system of divination, whereby the play of unseen forces is disclosed by God, the author of all power, through a visioner, in order that appropriate action could be taken. The efficacy of such a system rests on the assumption that all of creation, entities seen and unseen, are bound together in a web of spiritual energy, the texture of which can be influenced by access to *agbara Ẹmi Mimọ*. It is this empowered cosmology that is now losing its hold on the imagination of contemporary Yoruba in London.

As Keith Thomas (1971) has shown for post-Reformation England, the disenchantment of the world is a patchy business. With Born-Agains, as with Aladura, the extent to which the unseen world has become despiritualized varies between individuals, and at different moments in one person's experience. The C&S have a riposte to Pentecostal intransigence about Aladura practice. What about the Born-Agains who make clandestine visits to Aladura prophets? The extent of this is hard to verify, but it certainly occurs.

What seems clear is that despite vociferous assertions of abandoning previous traditions by both Aladura and Born-Agains, there lies a crooked thread of continuity. If, in looking back at Aladura through a Pentecostal lens, we focus on style and ritual, then we will indeed see rupture. However, if we include conceptions of spiritual power in the picture, we will see not only continuity, but how ideas of power are shaped, quite literally, by the "spirit" of the age.

Glossary

Agbara	Power
Agbara emi	Spiritual power
Agbara Emi Mimo	Power of the Holy Spirit
Aje	Witch, witchcraft
Aladura	Literally "praying." Generic term for Yoruba churches similar to the C&S
Ase	Power, spiritual power, authority
Babalawo	Diviner priest
Egungun	Masked ancestral cult
Elemi	Spiritualist; one who becomes possessed by the Holy Spirit
Emi	Spirit, breath, unseen being
Emi buruku	Bad spirit
Emere	Type of troublesome spirit
Epe	Curse
Ese	Verse from the Ifa divination text
Esu	Yoruba trickster God; the Christian Devil
Gelede	Masked performers/performances to honor witches and women's power
Ifa	Indigenous Yoruba oracle; the God of divination
Imo	Knowledge; intuition
Iran	Vision
Mimo	Holy, pure, clean
Oba	King, chief
Odu	Set of verses in the Ifa oracle
Ofo	Incantation
Ogun	God of iron and war
Okan	Soul, heart
Olodumare	God, "Almighty"
Olorisa	Priest, devotee of a deity
Olorun	God of Heaven

Ọmi mimọ	Holy water
Oniṣegun	Practitioner of indigenous "medicine"; herbalist
Oogun	Indigenous "medicine"
Oogun buruku/buburu	Harmful "medicine"
Oogun rere	Good "medicine"
Ọpa	Rod, staff
Oriki	Praise names
Oriṣa	Deity
Orukọ mimọ	Holy Name (of God)
Ọrun	Heaven
Orunmila	God of Ifa divination
Oṣo	Sorcerer
Ọta	Enemy
Ọya	Female deity of wind and tornadoes
Sango	God of thunder and fire
Sopona	God of smallpox

Notes

Chapter 1 Introduction

1. Studies published since 1980 include: Apter 1992; Barber 1991, 2000; Barnes 1986, 1997a; Berry 1985; Buckley 1985; Eades 1980; Drewal 1992; Drewal and Drewal 1990; Drewal and Mason 1998; Gleason 1987; Hallen and Sodipo 1986; Matory 1994; Peel 1983, 2000.

2. Of the few pieces that focus on Nigerian students the best is Fagbola 1960. The most informative general study on overseas students of this period is that by Morris and Ajijola 1967. For annotated bibliographies on overseas students in the late 1960s, see Kendall 1968 and Sivanandan 1969.

3. See Goody 1971; Goody and Muir 1972; Goody and Muir Groothues 1977, 1979; and the collection edited by Ellis 1978.

4. The C&S is also known as the K&S, from the Yoruba *Keruba ati Serafu*.

5. In what follows, I often use Aladura interchangeably with the C&S where appropriate.

6. See chapter 3, note 1.

7. Kalilombe 1997: 322.

8. Booth 1984.

9. Hunt 1973; Berner 1990.

10. See chapter 9.

11. For example, Barker 1983, 1992; Robbins 1988; Wilson 1990.

12. As by Hackett 1987a; Mbon 1991; Turner 1979.

13. See Peel 1993.

14. Gerloff 1999.

15. For Black Power ideology see Carmichael and Hamilton 1968.

16. Such as *The Sixth and Seventh Books of Moses*. See Chapter 8, pp. 197–98.

17. Buckley (1985) and Hallen and Sodipo (1986) also stress their collaboration with Yoruba ritual adepts.

18. For example, Durkheim 1964 and Marett 1932.

19. For example, by Ilesanmi 1983 and Gbadegesin 1991.

20. Evans-Pritchard 1936 and Seligman 1932: 522.

21. Ibid., 271.
22. Ibid., 167, 157; Seligman 1932: 525.
23. Ibid., 82.
24. Ibid., 11.
25. Ibid., 321.
26. Ibid., 33–34.
27. Ibid., 36.
28. Ibid., 26.
29. Ibid., 291.
30. Ibid., 219.
31. Ibid., 209, also 231.
32. Ibid., 21.
33. Danforth (1989: esp. chap. 8) and Lurhmann (1989) both analyze New Age ideology.
34. As by Marx and Engels 1957 and Bloch 1984.
35. Durkheim (1964), Malinowski (1954), and Weber (1966), from different perspectives, upheld this distinction, which Evans-Pritchard opposed (1965).
36. For example, Devisch 1985: 51–54 and Zeitlyn 1987: 40ff., 1990: 659.
37. Peek 1991b. Examples of this approach include: Beattie 1966, 1969; Beattie and Middleton 1969b; Bourguignon 1968; Devisch 1991. Shamanic practice also incorporates divination, as illustrated by Butt's material on the Akawaio in Wavell et al. 1966: 38–61, 83–105, 153–206. The equivalence is also illustrated in those societies where mediums coexist with the manipulation of material paraphernalia as an alternative method of divination (Beattie 1967b; Middleton 1969; Parkin 1991; Retel-Laurentin 1974; Whyte 1991).
38. As by Andersson 1968; Aquina 1967; Baëta 1962; Lee 1969; Sundkler 1961.
39. Jules-Rosette (1975, 1978, 1979), for example, argues that prophecy has a divinatory function, but gives almost no account of prophetic praxis.
40. For example, Moore 1957; Park 1963; Shelton 1965.
41. For example, Abbink 1993; Mendonsa 1982; Peek 1991a; Zeitlyn 1987, 1990.
42. See Davies 1999; Hastrup and Hervik 1994; Okely 1994, 1996a; Okely and Callaway 1992.
43. It is interesting that Edith Turner's fascinating account of her experience among the Ndembu (1987) is still completely overshadowed by the work of her husband, Victor Turner (e.g., 1967, 1975).
44. Anthropologists have indeed worked in the service of the state, both with British colonialism and U.S. counterinsurgency in Africa and elsewhere (Asad 1973; Stocking 1991).
45. The Redeemed Church of Christ (Miracle Praying Band) was led by Prophet A.O. Ajasa; the Primate of the Church of the Lord was Adeleke Adejobi; Aladura International is still headed by Father Olu Abiola.

46. The Kingsway International Christian Centre (KICC), the Glory Bible Church, the Kensington Temple, the Universal Church of the Kingdom of God (UCKG).
47. See Chapter 4, note 28.

Chapter 2 Yoruba Worker-Students in Britian

1. Information on income levels is taken from National Manpower Board studies (Federal Ministry of Economic Development 1965 and 1967); Federal Ministry of Information 1968; Teriba and Phillips 1971.
2. Sources on the nature of the Yoruba elite during this period include: Abernethy 1969; Ayandele 1974; Barnes 1986; Berry 1985; Lloyd 1966a, b, 1967, 1974; Marris 1961; Osoba 1977; Plotnicov 1970; Smythe and Smythe 1960.
3. See Marris 1961: 133 and Okediji 1967: 70–71. As in all strata of Nigerian society, networking is an essential elite strategy in employment and politics (Eades 1980: 157).
4. Coleman 1958: 134. The best general review of Nigerian education is found in Abernethy 1969.
5. Arikpo 1967: 111. For higher and university education in the first years of Independence, see Fafunwa 1971 and Okafor 1971.
6. The 1960 Ashby Commission estimated a need for an extra 54,700 workers with 2–3 years of higher education, including some 5,000 engineers, plus a possible further 31,000 with a degree or equivalent by the year 1970 (Federal Ministry of Education 1960). A market survey of the following year indicated vacancies for 1,000 skilled workers, including nearly 600 architects, engineers, and surveyors (Post 1968: 153).
7. Williams 1970, 1976.
8. Two-third of Smythe and Smythe's elite sample had qualified in England (1960: 75–76) as had 68% of the professionals in the 1965 Nigerian Manpower Survey.
9. In *The Edifice*, the narrator, a Yoruba student adrift in London, imagines such a scenario with dread (Omotoso 1971: 59–61).
10. From the Home Office, Control of Immigration Statistics: 1962–63 and 1967.
11. Great Britain, General Register Office, Summary Tables 1967: 35; Office of Population Censuses and Surveys, Country of Birth Tables 1974: 83. In 1971, there were another 10,000 Nigerians in Great Britain as a whole: 28,570. Statistics on Nigerians also include the communities of ex-seamen, who settled in the dock areas after the war.
12. British Council 1970: Statistical Supplement. Also see Muir and Goody 1970: appendix C, p. 7, table 14. I am grateful to Esther Goody for access to this unpublished report.
13. Harris 1972.

14. In her *Invention of Women*, Oyeronke Oyewumi (1997) argues that inequality between the sexes was a creation of colonialism, a theory challenged by Peel (2002), who provides evidence from nineteenth-century mission records of the gendered nature of Yoruba cultural life. For contemporary relationships between men and women, see Cornwall 1996.

15. In the 1972 Matrimonial Proceedings (Polygamous Marriages) Act: chap. 38 recognized polygynous marriages in Britain for those whose culture permits it.

16. Although Okediji (1967) argued that the middle class had smaller families, this is contradicted by the findings of Levine et al. (1967); Lloyd (1970); Lloyd (1967: 142ff.); Olusanya (1967).

17. The chief sources on West African fostering in London during this period include Ellis 1971, 1978; Goody 1971; Goody and Muir 1972; Goody and Groothues 1977, 1979; Muir and Goody 1970, 1972; UKOSA 1972.

18. Rose 1969: 134

19. See Muir and Goody 1970: appendix C, p. 5, table 11; Rose 1969: chap. 12; UKCOSA 1972: 16–17.

20. Daniel 1968: chap. 9. In his 1970 survey of Upper Holloway, where some C&S members lived, Rowland (1973: 63) found that "blacks pay over thirty per cent more for approximately half the rooms that the whites get."

21. Rose 1969: 133, 137. Out of the UKCOSA sample of 61 Nigerians, only one was a council tenant (1972: 16).

22. Weinreich 1986: 307.

23. Toulis 1997: 274; Wallman 1978: 207.

24. This is not to say that all African-Caribbean immigrants came from a proletarian background. Many were subject to the same stereotyping as Yoruba, which also resulted in occupational downgrading in Britain (Toulis 1997: 192).

25. In his analysis of the incorporation of black music and style into British youth culture, Gilroy (1987) does not mention West Africa.

26. The student in Omotoso's novel suffers the same scrutiny (1971: 55).

27. The literature on race in the 1960s and 1970s is extensive and includes Bourne 1980; Centre for Contemporary Cultural Studies 1982; Daniel 1968; Deakin 1970; Fryer 1984; Gilroy 1987; Hall 1978; Jenkins 1971; Moore 1975; Rex and Mason 1986; Sivanandan 1976, 1983; Smith 1977.

28. Smith 1977: 21.

29. For a contemporary analysis of Powellism and racism see Foot 1965 and 1969.

Chapter 3 The C&S Church, U.K.

1. Since Peel's initial study of Aladura in 1968, there has been an increasing interest in the development of Nigerian religious independency. Among

the most significant studies of the C&S on which I have drawn are those by Omoyajowo (1978, 1982, 1984), which include detailed accounts of C&S history. See also Olayiwola 1987; Probst 1989; Ray 1993. Earlier sources include: Mitchell 1963; *Nigeria Magazine* 1957; Parrinder 1951; and Turner 1965. As in Peel (1968), the history of C&S should be understood in conjunction with that of other Aladura. Harold Turner's two-volume *African Independent Church* (1967) remains the most thorough study of the Church of the Lord. See also his collected articles on *Religious Innovation in Africa* (1979). For a relative newcomer on the Aladura scene, the Celestial Church of Christ, see Crumbley (1992); Hackett 1987c; and Olupona 1987. Another is The Brotherhood of the Cross and Star (Mbon 1992). See also Hackett 1987b, 1991 and Mbon 1987, 1991 for brief overviews of Aladura in contemporary Nigeria. My own understanding of C&S history has been supplemented by C&S elders in Britain, and by interviews with Baba Aladuras Kalejaiye and Diya, and Captain Abiodun during their visit to London in September 1971.

2. For the history of the C&S in the eastern regions, see Moller 1968; Oguntomilade 1987; Omoyajowo 1982: chap. 5.

3. Oguntomilade 1987; Omoyajowo 1982, 1984; Peel 1968.

4. Omoyajowo 1982: 65.

5. Plotnicov 1967.

6. Little has been written on the C&S in the north, apart from Omoyajowo 1984: chap. 4.

7. Ogundele was speaking in tongues, which Fakeye interpreted (see chapter 6).

8. C&S Papers: letter to C&S Society, Okoko, Nigeria (n.d.).

9. For men: The Soldiers of Christ, the Light of Morning Star, Fogo (the Band of Glory), and Love Divine Fellowship. For women: Queen Sheba, Queen Esther, Mary, and Martha. There was also a group for children: Emmanuel Band.

10. Omotoso 1971: 33.

11. *The Times* 13 September 1968. The mental health of Nigerian students during the 1960s and 1970s was a cause for concern (Anumonye 1970: 16ff.; Fagbola 1960: 22; Lambo 1960; Morris and Ajijola 1967: 65; Rack 1982: 213).

12. Enquiry as to the "home town" of members revealed a diversity of Yoruba ethnic subgroups, tempered by a strong preponderance of Egba and Ijebu, both areas of early Western education.

13. These figures are based on a sample of male elders, and the members of two male Bands.

14. Barret (1968), Brandel-Syrier (1962), and Sundkler (1961) rightly stressed the new openings offered to women in African independency, but scholars writing after concepts of gender had become part of the academic canon note the limitations in women's role. See Crumbley 1992; Hackett 1987d: 203; Ludwar-Ene 1991: 67; Olupona 1987: 62.

15. For Biblical injunctions on menstrual pollution, see Lev. 15:19–33.
16. Judg. 16; 1 Cor. 5.

Chapter 4 The Concept of Spiritual Power

1. The continuities between indigenous practice and early Yoruba Christianity in the search for empowerment are explored in Peel 2000: esp. 255–265.
2. For example: Hackett 1983; Peel 1968; Ray 1993; and Turner 1967. Mbon (1992) makes the same point in his study of the Nigerian Brotherhood of the Cross and Star; Baëta (1962) and Beckman (1975) emphasize the desire for spiritual efficacy in the growth of Independence in Ghana; Kiernan (1990) and Comaroff (1985) stress the pursuit of spiritual energies by South African Zionists.
3. Such as Dennett 1968; Farrow 1926; and Meek 1943.
4. Buckley 1985: 9, 45; Drewal and Drewal 1990: 76.
5. Following from this, the word *aṣe* is used in C&S ritual for what is known as the "seal," the formula that is used to conclude prayer, ensuring its efficacy:

> Oba Alaṣe L'aiye ati L'ọrun
> Oba Alaṣe L'ọrun ati L'aiye
> F'aṣe si adura wa.
> Lai L'oruko Jesu Kristi Oluwa wa. Amin.

The translation used when the prayer is said in English runs as follows:

> O God that has power to seal both earth and heaven.
> O God that has power to seal both heaven and earth,
> Come and seal our prayer.
> In the name of Jesus Christ our Lord. Amen.

6. For example: Rom. 13:1–3; 1 Cor. 15:24; Luke 10:19.
7. Judg. 16.
8. Luke 1:51.
9. For example: Jer. 10:12; Rom. 1:16, 19–20; 1 Cor. 4:20; Eph. 1:19.
10. Pss. 59:11, 16; 66:3; 110:3.
11. For example: Acts 8:10, 26:18; 2 Thess. 2:9.
12. Matt. 10:1; Mark 3:15; Luke 4:36, 5:17, 9:1.
13. Acts 3:12, 10:38.
14. John 17:2; Eph. 1:19.
15. Ps. 78:26.
16. Eccles. 8:4; Luke 4:32.
17. 1 Cor. 2:4–5; Mic. 3:8; Eph. 3:7, 20; Rev. passim.
18. Rev. 11:3, 6.
19. Luke 1:35.
20. Matt. 24:30; Luke 21:27.
21. Acts 1:8; 2.
22. Matt. 6:13.
23. Gleason 1987.

24. Drewal and Drewal 1990.
25. McKenzie 1976.
26. The orthodox definition of a sacrament as "an outward and visible sign of an inward and spiritual grace" is quite consistent with the idea of ritual transference of spiritual power.
27. Giddens 1976: 53, 111–113.
28. Irele (1975: 88) points to the "lively reciprocity" between the "everyday" and the "extraordinary" in classic Nigerian fiction by D.O. Fagunwa, Amos Tutuola, and Wole Soyinka. The same theme also informs the mindset of characters in popular novels such as those by T.M. Aluko (1966, 1970), Lakan Oyegoke (1984), Jide Oguntoye (1987), and 'Sola Osofisan (1991). It is a leitmotif of autobiography such as that of Augustus Adebayo (1988) and Wole Soyinka (1983).
29. *Oluwa awọn Ọmọ Ogun*. Here the Prophet emphasizes the combative aspect of the "Lord God of Hosts."
30. Hence the importance for Aladura of Isa. 11:2: "And the spirit (ẹmi) of the Lord shall rest upon him, the spirit of wisdom and understanding, the spirit of counsel and might, the spirit of knowledge and of the fear of the Lord."
31. 1 Cor. 12:1–11.
32. Here we have *aṣẹ* used in the sense of God's authority that carries nuances of the indigenous concept, but is glossed by the C&S translator as "seal" rather than "power."
33. On Yoruba witchcraft, see: Apter 1992; Awolalu 1970, 1979; Beier 1958; Drewal and Drewal 1990; Hallen and Sodipo 1986; Idowu 1962, 1970; Lucas 1948: 283ff.; Morton-Williams 1956a, 1960a, 1967; Osunwole 1991; Parrinder 1951: 53ff.; Prince 1961, 1964; Thompson 1970.
34. Gleason 1987: 300. Also see Drewal and Drewal 1990: 75.
35. Prince 1961: 796.
36. A ritual staff signifying the prophet's spiritual authority.
37. See Peel 2002 on detailed evidence drawn from nineteenth-century CMS sources.
38. Drewal 1992; Drewal and Drewal 1990; Gleason 1987; Hallen and Sodipo 1986.
39. Aluko 1966, 1970; Barber 2000; Oguntoye 1987; Oyegoke 1984.
40. This distinction, first elaborated by Evans-Pritchard (1937), has influenced interpretations of Yoruba witchcraft such as that by Awolalu (1979: 91). For a critique of this position, cf. Hallen and Sodipo (1986).
41. Although there is little on *emere* in the literature, these spirits are another part of the traditional Yoruba cosmology that is carried over into the Aladura universe. Afflicting women, *emere* is said to be a "sister to witchcraft," only less powerful. Members describe *emere* as a spirit that troubles the ẹmi of the invaded woman herself, rather than referring to the power of an aggressor. Although she will have less control over the spirit than in the case of witchcraft, it also may be able

to provide a woman with things she wants, including revelatory power; however, *emere* also causes the deaths of *abiku*, children who are repeatedly called back to Heaven in a cycle of death and rebirth. Emere spirits are said to manifest themselves in possession, and elders privately cite cases of women in the church who are harassed by them.

42. Oyetade, on the universal belief in ọta (n.p., ca. 1996: 1), lists 18 different categories of enemy identified by Yoruba (3–4). These reveal an especial mistrust of those who mask malice behind feigned friendship—as in the warning I received "through dream" when I first attended the church (see chapter 1, p. 7).

43. The majority of hymns were sung in Yoruba from *Iwe Orin Mimọ* (Holy Hymn Book). The ESO. Holy Hymn Book was the official English version of the C&S Hymnology, and included both translation of Yoruba texts and English CMS hymns.

44. Lambo (1960: 101, 1963: 8, 1964, 445) and Stone (1965, 1969).

45. Willis (1968) found that educated Fipa were moving toward attributing sickness to physical poisoning, rather than supernatural agents as before.

Chapter 5 Dynamic Metaphors of Spiritual Power

1. Yoruba fiction reflects popular perceptions on this issue, i.e., Aluko 1966: 191. Novelists often represent *oogun* and "spiritual" churches as functional alternatives: Aluko 1966: 125; Oguntoye 1987.

2. On interest in the power of *orisa* and ancestors in the mid-twentieth century, also see Awolalu 1970: 33; Buckley 1985: 116; Davis: 1977; Morton-Williams 1956b: 90, 1960b: 373.

3. Deut. 18:10–12; 2 Chron. 33:6.

4. Cf. Pss. 115:4–7; 135:15–17.

5. Anglican palm-leaf crosses commemorate Christ's entry into Jerusalem (John 12:13) and celebrate the resurrection as well as the crucifixion.

6. Drewal 1989: 257; Drewal and Drewal 1990: 75; Gleason 1987: 96–97. See also Buckley 1975: chap. 6. Palm fronds have wide cross-cultural currency as veils of the sacred.

7. Peel 2000: 217–218.

8. Peel 1978: 144.

9. Gen. 1:1–5.

10. For example: Ps. 27:1; Matt. 5:14, 16; John 1:9, 12:34–36; 1 John 1:5.

11. Exod. 3.

12. Acts 9:3–9, 22:6–11, 17–21.

13. Rev. 4:4.

14. On the significance Yoruba attribute to the color red, cf. Buckley 1985: chaps. 3 and 4; Drewal and Drewal 1990.

15. Exod. 13:21–22.

16. Wescott and Morton-Williams 1962: 323.

17. Barnes 1997b: 18. Ilesanmi (1993: 64) argues that the binary expression of Yoruba language deals with complementarity rather than opposition, exemplified by the character of divinities.

18. This doctrine on the nature of power is neither archaic nor confined to ritual specialists; Aladura are in tune with the discourse of contemporary Yoruba fiction. See Oyegoke 1984: 42–43.

19. That is, a member of the *Egungun* ancestor cult. See Beier 1956a; Davis 1977; Morton-Williams 1956b.

20. Lawuyi (1986 and 1988: 10–11) argues that the ubiquitous references to God in popular discourse fuse indigenous with Christian notions of the source of power. This continuity in the conception of God emerges in the writing of D.O. Fagunwa. Take this excerpt from a character's prayer, quoted in Olabimtan (1975: 106):

    All men pray to you, You answer their prayers. Christians pray to You, You answer their prayers. Muslims pray to You, You answer their prayers. Sango worshippers pray to You, You answer their prayers. Oya worshippers pray to You, You answer their prayers. All worshippers of Orisa pray to You, You answer their prayers. Why is it, my God, my King, my Saviour that You will not answer my own prayer?

21. This principle is vividly dramatized in Amos Tutuola's *The Wild Hunter in the Bush of Ghosts*. Charms, witchcraft, ghosts, humans, God, and the Devil—"every kind of supernatural power" (1989: 84) pit their strength against each other in a wonderful dreamlike allegory of comparative strength.

22. Morton-Williams 1960a: 35. See the formula quoted in Maclean 1971: 95.

23. Representing aggressive power, see 159.

24. This tale is reminiscent of an incident in Oyegoke's *Laughing Shadows*; a wife was preparing a medicine to kill her husband, when "the kitchen exploded in her face" (1984: 192).

25. This vision echoes the parable of the talents: Matt. 25:14–30.

26. Goody 1993: 11–23.

27. The prophet who translated this vision for me confirmed that the images were referring to the brother's spiritual power.

28. Bicycles, motorcycles, and airplanes feature on masks worn in Gelede rituals to honor the power of "the mothers." See Drewal and Drewal 1990: 201–203.

29. The clearest examples of this deduction from encounters with airplanes are the cargo cults of Melanesia (Burridge 1960; Lawrence 1964; Worsley 1957). Stoller (1995: 83–90) reads the mimicry of the colonizers and their material goods in West African plastic arts as an attempt to co-opt the white man's power.

30. Fernandez 1990.

31. See also 13:1 and 19:2.

32. Beier 1982; Drewal and Mason 1998; Thompson 1970.

33. "Only members from the grade of Evangelist upwards have the right to use staff. As the use of staff signifies power, those entitled to use it must ensure that proper use is made of such staff and that they [*sic*] are not used for purposes other than those in the interest and progress of the church. The Executive Committee reserves the right to withdraw staff where a case of wrongful use is ascertained" (General Codes of Conduct. n.d. Section VI: 4).
34. Rev. 2:27.
35. Exod. 7 and 8.
36. In Ps. 110, enemies are routed by the "rod of thy strength" (*ọpa agbara*. v. 2), "in the day of thy power" (*ọjọ agbara*. v. 3).
37. Transformation is another theme in Amos Tutuola's tale of *The Wild Hunter*. So, for example, the Hunter turns night into day with the "miraculous power" of his spell, when fighting a ghost who could make the day become night (1989: 35–36).
38. Exod. 4:2.
39. 1 Sam. 14:27.
40. Deut. 34:9.
41. 2 Kings 2:9–15.
42. Morton-Williams 1967: 322, 1972.

Chapter 6 Possession by the Holy Spirit

1. In Nigeria, Watchnight services begin at midnight, and may last from four to six hours. In London, they were held every month from 5 September 1970, after the purchase of Earlham Grove, and I attended six. This account is based on the service of 9 January 1971.
2. For an evocative account of dancing to the internal rhythms of possession, see Fatima Mernissi (1995: 172) on the Moroccan *hadra* possession dance.
3. The most thorough analysis of glossolalia to date is that by Goodman (1972: see 149–151). Abasiattai (1989: 505) argues that the sounds and inflections produced in trance may be similar to those of languages that the glossolalist has heard spoken.
4. An example of a Watchnight consecration prayer is given in chapter 4.
5. This is a translated summary (by Korode himself) of the prayer, which was longer and more repetitious.
6. On 2 May 1971, for "Nationwide."
7. The understanding of possession as penetration of the human body by a spirit, together with the seemingly erotic movements that accompany trance, result in an explicitly sexual vocabulary of possession in many cultural contexts (Danforth 1989: 81, 88; Lewis 1971: 59–64). Both in Yoruba traditional ritual and versions of this in the diaspora, such as in Haiti and Brazil, the spirit is said to "mount" his "bride" (Deren 1953; Matory 1994). But there are no sexual references in

C&S trance, which does not, despite appearances, involve sexual arousal or orgasmic release.

8. Calley 1962, 1965.
9. Wilson 1961.
10. For example: Boddy 1989: 147–150, 256; Crapanzano 1977: 14; Lambek 1981: 5, 49–53; Lewis 1971: chap. 7; Macklin 1977: 66–74; Stoller 1989a.
11. Inglis 1990: 221–222; Jilek 1989: 182; Valla and Prince 1989.
12. Chevreau 1994.
13. Howard 1996.
14. That is, by Poewe, who defines charismatic Christianity as that which has "emphasised religious or spiritual experience and the activities of the Holy Spirit" (1994: 2).
15. For example, by Inglis 1990 and Jilek 1989.
16. The ability of particular culturally conditioned cues to penetrate trance is well documented. See, for example, Lambek's account of the "smooth synchronization" of the movement among a cast of spirits as they switch from one host to another, never possessing two people at once. An observer sees chaos and confusion, but the whole event is carefully choreographed through subtle signals sensed by the entranced (1981: 110–114).
17. C&S prophets hold that there are degrees of dissociation. One claimed that vision would inform him about someone in Spirit "that he is on the third ladder (i.e., rung), and has got four more to climb."
18. But lengthy periods of possession in London may well take place. The Observer (23 January 1994) reported on a C&S prophetess who was 16 days into a trance, which she had prophesied would last for three weeks.
19. See Lambek (1989) for a concise statement of this position, which informs all recent research on spirit possession.
20. When young East African Samburu and Masai men succumb to trance, for example, it is thought merely to express anger and frustration (Lewis 1971: 40; Melissa Llewellyn-Davis: personal communication).
21. For example, Lambek 1981.
22. For example, Boddy 1989.
23. Both this and the following quotations refer to the feeling of empowerment against enemies that the Spirit brings.
24. A few observers have attempted to convey the power of the event through their own experience (e.g., Deren 1953; Dunham 1994; Goodman 1972: 72). Boddy herself confesses, "At some zar rituals I became apprehensive lest I succumb to trance and lose self-control" (1989: 349). As Desjarlais (1992: 14–19) warns, the ethnographer's own encounter with trance may not be a reliable guide to indigenous experience, but Boddy's fears reveal an overprivileging of textual analysis to the near-exclusion of the experiential.
25. Bourguignon and Pettay 1964: 40.
26. See Ward 1989: 16–19.

27. See, for example, Lutz and Abu-Lughod 1990.
28. To what extent visioning depends on trance remains unclear. Both the Spiritual Leader and the Pastor, neither of them spiritualists, claimed to have seen visions, but the consensus among prophets is that visioning develops out of dreaming and dissociation. This is borne out by most individuals' accounts of their spiritual careers, where falling in Spirit preceded their ability to vision. Once the response is well established in an experienced visioner, revelation becomes a natural part of prayer, and dissociation may be very light.
29. See Omoyajowo (1982: 137–141) for similar practice in Nigerian C&S churches.
30. See 1 Cor. 14 for the scriptural foundation of this position.
31. Infusion with the power of the Holy Spirit is a widespread interpretation of trance, which features in many African independent religious movements (Barret 1968; Wilson 1973), whether South African Zionists (Comaroff 1985; Kiernan 1990; Sundkler 1961), Zimbabwean Apostles (Aquina 1967), Ibibio adherents of Oberi Okaime (Abasiattai 1989), or Ghanaian independents (Baëta 1962; Beckman 1975; Hagan 1988), to name but a few.
32. See also Num. 11:26; 1 Sam.19:20, 23; 1 Chron.12:18, 28; 2 Chron.15:1; Ezek. 2:2, 3:12.
33. Although Apostolic doctrine included "visionary guidance of the 'prophetic ministry' " (Peel 1968: 105), this was not central, and the role of revelation also appears to have been elaborated by Aladura themselves. Peel states that "reliance on visions for guidance . . . was a spontaneous development" (65), and Turner (1967 II: 137–140) reaches similar conclusions for the Church of the Lord.
34. Awolalu 1979: 37; Matory 1994: 175; Verger 1969.
35. Drewal 1992: 184.
36. Apter 1992: 106, 136; Beier 1956c.
37. Awolalu 1979: 37. See also Parrinder 1961: 73; Prince 1964: 114ff.
38. Apter 1992: 237 n.6; Beier 1956a.

Chapter 7 Revelation as Divinatory Practice

1. The choice of language is a personal one, which depends on the visioner's feeling of comfort and fluency in the language chosen. When visioners were clearly "in Spirit," they tended to speak in Yoruba. They also might switch to Yoruba from English when directing a message to an individual that did not concern the whole congregation. Yoruba visions that did address the group would be translated into English at a later moment.
2. As this sequence of visions has been abbreviated, numbering is for the purposes of reference, not to record the actual quantity of messages.
3. The move to Earlham Grove was imminent.

4. For the Aladura practice of dreaming, see also Omoyajowo 1982: 131–133; Peel 1968: 123, 170–171, 206–207; Turner 1967 II: 122–125. C&S pamphlets recommend ways of developing "the Spirit of dreams" and of gaining information through dreaming, especially through the use of psalms (i.e., Akindele n.d.; Sonde n.d.). A vision once instructed me to read Ps. 6 "*on* water" (i.e., *over* water, which should then be used as a drink or to wash my face). This psalm was usually recommended for eye trouble (see v. 7), but I was to use it for "the real power of dreams." Visioners often refer to the "spiritual eye" of both vision and dream.

5. See Maclean 1971: 93; McKenzie 1992; Prince 1964: 95.
6. See also Deut. 13:1, 3, 5; Job 33:15–16; Dan.7:1, 2, 7, 13, 15.
7. Gen. 37, 42:9.
8. Dan. 1:17, 2.
9. 1 Kings 3:5–15.
10.

Date	Type of service	No of visions
1 June 1969	Ordinary Sunday	40
18 January 1970	Ordinary Sunday	113
18 May 1969	Revival	66
1 February 1970	Revival	26
2 August 1970	Band Anniversary	59
6 September 1970	Band Anniversary	101
5 August 1970	Watchnight	142
7 November 1970	Watchnight	131
	Total	678

Quoted visions are taken from the above or selected from the hundreds of revelations recorded from a further 18 services.

11. Fiske 1989.
12. See also: #3, 7, 13, 16, 17, 18, 24, 44, 48, 49, 50, 51–53, 56, 65, 67–69, 70, 72.
13. See also: #11, 12, 27, 45, 46, 55, 57, 58.
14. See also: #19, 28, 45, 33, 39, 61, 64.
15. See also: #13, 24, and 30.
16. For example, Skultans 1971: 992.
17. Jung makes this explicit in his Foreword to the *I Ching* (Wilhelm 1968: xxxviii–xxxix).
18. The figures are 32% to men, 10% to women, and 1% with gender unspecified.
19. See Jules-Rosette 1978; Shaw 1991.
20. Boddy (1989: 135) argues that "the system of meaning . . . is both logically and contextually prior to the behaviour . . . through which it finds expression." See also Lambek 1989: 38.
21. Hollis and Lukes 1982.

22. Some earlier ethnographers, such as Beattie (1967b: 231) or Shelton (1965: 1444–1449), saw in the "irrational" methods of divination a compensation for faulty logic. But this positivist sin has been overemphasized. For example, whereas Devisch (1985: 66) takes Victor Turner to task on this score, Turner himself expressly states that diviners act in an "astute and rational way," given their convictions on "spirits and mystical forces" (1975: 229).

23. For a more detailed analysis of Ifa, see chapter 8. There is a wealth of literature on Yoruba oracles: both Bascom (1941, 1942, 1943, 1961, 1966, 1969, 1980) and Abimbola (1965, 1977, 1994) have had a lifelong interest in the subject. Idowu (1962) and Awolalu (1979) include divination in their overview of Yoruba belief, as does Parrinder (1951, 1961). Recent work includes Akinnaso 1995; Gleason 1973; Makinde 1983. For earlier studies see Clarke 1939; Dennett 1968; Lucas 1948; McClelland 1966.

24. Oguntoye 1987.

25. See Nassau 1904; Platt 1935: 15, 45–51.

26. See also: 2 Kings 17:17; Ezek. 13:23; Zech. 10:2; Acts 16:16.

27. 1 Sam. 9:6.

28. 2 Chron. 26:5.

29. 1 Sam. 23:2–4, 10–11; 2 Sam. 2:1, 5:19, 23. For the practice of prophecy among the Israelites, see Caquot 1968.

30. Gen. 15.

31. 1 Sam. 9:15–17.

32. 1 Kings 17:2–4, 8–9.

33. 2 Sam. 24:11.

34. Matt. 3:17, 17:1–9; John 12:28.

35. Acts 10:10–16.

36. Acts 9:36, 16:9, 18:9.

37. For example, Ananias and Cornelius (Acts 9:10–16; 10:3–6) and Zachariah (Luke 1).

38. For example, by Mendonsa 1982; the contributors to Peek 1991a; and Zeitlyn 1987, 1990.

39. Bascom 1969; Gleason 1973.

40. Abimbola 1994: 101; Bascom 1966, 1969: chap. 1; Palmié 1995. Recently there have also been attempts to simplify Yoruba divination for use in the Americas, such as by R.T. Kaser (1995) and Afolabi Epega (1985). Chief Kola Abiola's pamphlet on *Erin-dilogun* ("Sixteen Cowries") "is actually prepared for priest/priestess of Yoruba religion in the diaspora, who are anxious of continuous links with the *SOURCE*" (n.d.: 7).

41. The way this divinatory democracy affects the diviner rather than the client will be considered in chapter 8.

42. Shelton 1965: 1445–1447; V. Turner 1975: 209.

43. Akinnaso 1995. The ability of Yoruba ritual specialists to innovate and invent is the central theme of Margaret Drewal's *Yoruba Ritual*

(1992). The leeway for incorporating change has also been noted in the practice of other diviners, both those using spirits (Parkin 1991: 187–88) and those relying on mechanical systems of divination (Mendonsa 1982: 130).

Chapter 8 Revelation and the Power of Prayer

1. The information I asked the visioners to record about each consultant covered (a) The consultant(s): gender, church membership, where members had joined C&S—Nigeria or England, how often they attended Sunday services, whether the visioner had prayed for them before. (b) The consultation: date, location, whether the Prophet was alone or praying with others, the purpose of the prayer, visions revealed and advice offered. The prophet's name and certain other identifying details of both clients and visioners have been changed. I am extremely grateful to those who offered both trust and time in order to provide an insight into their activity.

2. Prophetic entrepreneurs outside the church are free from such control, such as the independent Apostle in West London who charged exorbitant prices for prayer and ritual paraphernalia, and specialized in holy baths for his female clientele. Abuse of clients may also occur inside the church. In 1992, the C&S received unfortunate publicity on the front page of the *Daily Sport* concerning a pastor prosecuted for alleged sexual harassment: "Randy Rev's Big Red Candle." The story was also reported—more soberly—in *The Times* (31 January 1992).

3. On personal messages delivered by spiritualists, see Nelson 1969 and Skultans 1971.

4. For sources on the Ifa oracle, see chapter 7, note 23. For other forms of Yoruba divination, see Bascom 1980; Idowu 1962: 135–137; Lucas 1948: 285; Parrinder 1961; Stevens 1965. Clients of all religious persuasions also seek the divinatory services of the Muslim *Alfa* (Abdul 1970; Bascom 1966, 1969: 8–10; Morton-Williams 1966: 407–408; Parrinder 1953: 80).

5. Popularization of Ifa for a wider audience, which is unfamiliar with the cosmological assumptions behind active divination, inevitably reduces its meaning. Whereas R.T. Kaser's *African Oracles in 10 Minutes* does mention the importance of "making a sacrifice and taking a remedy" (1995: 260), Afolabi Epega's booklet on Ifa offers brief instructions for the "person who desires to have his fortune told . . ." (1985: 35).

6. The withholding of the presenting problem from the diviner is a common feature of oracles. See Devisch 1991: 120; Mendonsa 1982: 119; Meyer 1991: 95; Parkin 1991: 177; Peek 1991c: 195; Whyte 1991: 165.

7. See Bascom 1941: 50–51, 1942: 42; McClelland 1966: 426. Evans-Pritchard (1937: 185–194) and Beattie (1967b: 219) record the testing of diviners by the Azande and Bunyoro.

8. That is, doubt God. Cf. John 20:24–29.

9. McClelland 1966: 427.

10. That is, as among the Azande (Evans-Pritchard 1937), the Sisala (Mendonsa 1982), the Nyole (Whyte 1991), and in the Bamum Kingdom, Cameroon (Tardits 1996). Significantly, in the private form of Temne divination, wrongdoers are not named—in contrast to witchfinding, "in which the calling of a public divination performance is in itself an acknowledgment that social relations have already been disrupted" (Shaw 1991: 147–148).

11. For example, see Bascom 1969: 149–153, 384–385, 528–529.

12. A brand of credit card.

13. For the equivalence of the Yoruba concept of sacrifice (*ebo*) and Aladura prayer as means for accessing spiritual power, see Peel 2000: 256, 265.

14. The same dual meaning appears in Northern Movement literature, *Ofin, Ilana ati Eto Egbẹ Mimọ Kerubu ati Serafu*: 50.

15. Barber 1991.

16. These meanings for candles, as for other ritual symbols, are contained in C&S pamphlet literature: Phillips 1962 and The Eternal Sacred Order of the C&S *The Order* . . . Also see Omoyajowo 1982: chap. 8. Turner (1967 II) finds a similar range of meaning in the Church of the Lord.

17. Hence the danger of infection by a sickness or evil spirit if using water consecrated for another.

18. These texts were much in vogue in West Africa (Field 1960; Turner 1967 II) and the Caribbean (Simpson 1956). Their historical origins appear to be a jumble of Solomonic legend, Egypto-Hellenic magic, Christian-Jewish mythology, Cabbalistic tradition, Gnostic and Arabic folklore, European occultism, the Talmud, and the Bible plus Apocrypha (Butler 1949; Sebald 1988). They have undoubtedly influenced Aladura (Peel 1968: 142; Turner 1967 II: 73–74; Wilson 1973: 182). In Britain, this literature, along with the requisite ritual props, were available from a mail-order firm, De Laurence, which, according to its publicity literature, had a flourishing trade in Nigeria.

19. The ritual instructions accompanying the use of Psalm 119 in *The Sixth and Seventh Books of Moses* include inscribing a hardboiled egg with particular Holy Names, which will then be eaten for good memory, desire to learn, and an "extended intelligence" (n.d.: 197).

20. Levit. 14; Exod. 26.

21. The Red Sea parted for Moses, said Ayorinde, "because of the operations he did" (cf. Exod. 14).

22. Such as the spittle and clay to anoint the blind man's eyes (John 9:6).

23. Exod. 12:7 describes the response of angels to "signs."

24. See also Jer. 23:32 condemning "them that prophesy false dreams." The Old Testament is full of stories of the untoward results of false vision; the New Testament of warnings against them (Matt. 24:11, 24; Mark 13:22).

25. Compare this with Okely's comment on the fakery of gypsy fortune-telling: gypsies never consulted each other (1996b: 100).
26. This interaction in Nyole divination is analyzed by Whyte (1991). See also Abbink 1993; Mendonsa 1982; Parkin 1991; Peek 1991c.
27. McClelland 1966: 421; Barber 1991: 103, 288–290, 327–328.
28. Burridge 1979: 192. For a discussion of ritual, both secular and sacred, as symbolic capital, see Bourdieu 1977: chap. 4, and Harrison 1992.
29. Report of the Spiritual Committee 2 November 1969.

Chapter 9 Empowerment and Yoruba Christianity

1. Karl Maier's journalistic account (2000) provides an excellent picture of recent Nigerian history.
2. Daley 1996: 47.
3. Ibid., 47, 49.
4. Olukoga 2001.
5. Storkey 2000.
6. Daley 1996: 53.
7. This was the theme of Rachael Rendall's documentary *Black on Black*, Juniper Productions for Channel 4, 23 September 2001.
8. Known as "419" after a relevant clause in the Nigerian criminal code. This is not the street crime commonly associated with African-Caribbean youth, but large-scale scams ranging from benefit fraud, drug-running, and mail-shots persuading the gullible to part with their money through promises of fictional profits.
9. See Matt. 10:16.
10. Sources for the recent history of the church include C&S pamphlets (Abidoye 2001; C&S Council of Churches 1990; C&S Church Movement 2000; Chirwa 1990); attendance at church services (C&S Earlham Grove, C&S ESO Mount Zion; C&S ESO Thursday Prayerist Church, New Covenant C&S, C&S Church Movement No. 6); participation in the School of Theology as board member and lecturer; meetings of the C&S Council of Churches; interviews with church leaders and other C&S members.
11. Although the C&S is the largest Aladura church in Britain, the first branch of the Church of the Lord was founded a year earlier, in 1964 (Berner 1990; Hunt 1973) and now has several branches in Britain. Harold Turner's two-volume study on the Nigerian church (1967) is a classic of Aladura literature. At the time of my original fieldwork, there were several other smaller Aladura congregations. The Redeemed Church of Christ (Miracle Praying Band), formed in 1967 by the exotic figure Prophet A.O. Ajasa, has since disappeared, as has Odonko's Universal Prayer Fellowship, and Akinwowo's 1962 Rebirth Mystical Order. Newer on the scene is the Celestial Church of Christ (CCC)

(Kerridge 1995), a Yoruba Aladura church that sets much store by the performative power of ritual symbol. There is as yet no detailed study of the CCC in London, although this "white-garment" church was the subject of a sensationalist television documentary (*White Church, Black Magic*, BBC2, 12 December 1994). For the Nigerian CCC, see Carter 1997; Crumbley 1992; Hackett 1987c; Olupona 1987. There have also been a number of other West African indigenous churches with Aladura praxis. Buafo's Universal Prayer Fellowship, founded in 1967, merged with an offshoot of the Church of the Lord (Gerloff 1992, Vol. 1: 247) so is no longer extant, but the Ghanian Brotherhood of Cross and Star, well established in Nigeria (Mbon 1987, 1991, 1992; Offiong 1987), has also attracted Yoruba members in London (Kerridge 1995).

12. Coker's involvement in ESO churches in London is recorded in the hagiography by Oguntomilade (1987: chap. 11).

13. The second broad grouping of ESO churches (including C&S Society and Mount Zion congregations), set up similar structures in 1999, which include branches in Germany and Holland.

14. Booth 1984.

15. There is a corresponding growth of literature. I have chiefly relied here on Coleman 2000; Corten and Marshall-Fratani 2001; Cox 1996; Gifford 1998; Martin 1990, 2002.

16. For Pentecostal influence on Aladura, see Peel 1968: 105, 108, 146–148, 298. For the Nigerian Born-Again movement, see Hackett 1995, 1998; Ojo 1987, 1988, 1995; Marshall 1993, 1995; Marshall-Fratani 1998. Although I have drawn extensively on these sources, it is important not to read off British Born-Again experience from that in Nigeria.

17. Ojo 1987: 17.

18. Gifford 2001.

19. For Nigerian Born-Agains in London, see Hunt and Lightly 2001. I have also drawn on an unpublished paper by Ojo (1997), but have mainly relied on the literature, services, and meetings of KICC, KT, and the Glory House, together with interviews with members and pastors.

20. Hunt and Lightly 2001: 109.

21. Freston 2001; Oro and Seman 2001. Also see the UCKG publication *City News*.

22. The impact of media technology features prominently in recent studies of Pentecostalism and is the frame for some of the most suggestive material on Nigerian Born-Agains, such as Hackett 1998 and Marshall-Fratani 1998.

23. Probst 1989.

24. The Rev. Mrs. Faith Oyedepo is the wife of Bishop David Oyedepo, now the head of the Winners Chapel, a 50,000-seater auditorium in Lagos, and author of several books sold on the Born-Again circuit. As head of David Oyedepo Ministries International, his travels include ministry at the London Glory House. Although Yoruba-headed

churches in London are independent of Nigerian foundations, there is a strong international Born-Again network centering on the leading preachers. This is familiar from U.S. Pentecostal practice.

25. See Toulis 1997 for a similar analysis of the African-Caribbean New Testament Church of God.

26. Burridge 1979; Dumont 1985; Lukes 1973.

27. Morris argues that Yoruba society is not characterized by the "all-inclusive collectivism" of many other African peoples (1994: 138). For the effect of the nineteenth-century Yoruba wars in fostering a spirit of regional and individual rivalry, see Lloyd 1974: chap. 2; Matory 1994: chap. 1. Beier (1982) and Lawuyi (1986: 110–111) also point to the space for individual aspirations and popular participation in traditional political life. The hereditary principle partially determined status, but political office rested also on election and popular support. Economically, both men and women had room to rise. In Yoruba theology, destiny is God-given, yet open to revision; ritual and cultural life thrives on innovation and individual preference (Barber 2000; Drewal 1992; Drewal and Drewal 1990: 105).

28. For a discussion of this point, see La Fontaine 1985 and Abrahams 1986: 56.

29. See Eames 1992 for a humorous but suggestive account of an anthropologist's own encounter with this culture.

30. Marshall 1993, 1995; Marshall-Fratani 1998; Peel 2000: 315.

31. Without further research, it would be dangerous to argue for a wholesale shift away from notions of external causation by younger men and women. The Celestial Church of Christ, which caters for the more magical end of the Aladura market, is still strong in London, whilst showing remarkable growth in Nigeria at the end of the twentieth century (Carter 1997: 76).

32. Wilson 1985, 1990.

33. Hammond 1985; Pro Mundi Vita 1989; Richardson 1985.

34. In 1970, 22% of the population was members of a Christian church; by the end of the twentieth century this had nearly halved to 12% (Brierly 1999). The Anglican church is still established—but on Easter Day 2001, less than 2% of the population went to church.

35. In 1991, 72% claimed to believe in some form of supernatural power (Bruce 1996: 33).

This page intentionally left blank

Bibliography

Abasiattai, Monday B. 1989. "The Oberi Okaime Christian Mission: towards a history of an Ibibio Independent Church." *Africa* 59:4, 496–516.

Abbink, J. 1993. "Reading the entrails: analysis of an African divination discourse." *Man* 28, 705–726.

Abdul, M.O.A. 1970. "Yoruba divination and Islam." *Orita* 3:1, 17–25.

Abernethy, David B. 1969. *The Political Dilemma of Popular Education.* Stanford: Stanford University Press.

Abimbola, Wande. 1965. "Yoruba oral literature: Ifa." *African Notes* 2:3, 12–16.

———— 1977. *Ifa Divination Poetry.* New York: Nok Publishers.

———— 1994. "Ifa: a West African cosmological system," in *Religion in Africa: Experience and Expression,* ed. T. Blakeley, W. van Beek, and D. Thompson, 101–116. London: James Curry.

Abiola, Kola. [n.d.]. *Ẹrin-dilogun.* Ile-Ife: Ife Cultural Centre Publications.

Abraham, R.C. 1958. *Dictionary of Modern Yoruba.* London: University of London Press.

Abrahams, Roger D. 1986. "Ordinary and extraordinary experience," in *The Anthropology of Experience,* ed. Victor Turner and Edward Bruner, 45–72. Chicago: University of Illinois Press.

Adebayo, Augustinus. 1988. *Once upon a Village.* Nigeria: Heinemann Educational Books.

Adesanya, A. 1958. "Yoruba metaphysical thinking." *Odu* 5, 36–41.

Ajayi, J.F. Ade. 1963. "The development of secondary grammar school education in Nigeria." *Journal of the Historical Society of Nigeria* 2:4, 517–535.

Akeredolu-Ale, E.O. 1971. "Values, motivations and history in the development of private indigenous entrepreneurship. Lessons from Nigeria's experience, 1946–1966." *Nigerian Journal of Economic and Social Studies* 13:2, 195–219.

Akinnaso, F. Niyi. 1995. "Bourdieu and the diviner: knowledge and symbolic power in Yoruba divination," in *The Pursuit of Certainty: Religious and Cultural Formulations,* ed. W. James, 234–257. London: Routledge.

Akinrinsola, Fola. 1965. "Ogun festival." *Nigeria Magazine* 85, 84–95.

Akintunde, J.O. 1974. "Nigerian national character: a political science perspective." *Odu* 9, 95–114.

Aluko, T.M. 1966. *Kinsman and Foreman*. London: Heinemann.
——— 1970. *Chief the Honourable Minister*. London: Heinemann.
Andersson, E. 1968. *Churches at the Grass Roots*. London: Lutterworth Press.
Animashawun, G.K. 1963. "African students in Britain." *Race* 5:1, 38–47.
Anumonye, Amechi. 1970. *African Students in Alien Cultures*. Buffalo, NY: Black Academy Press Inc.
Appadurai, Arjun. 1996. *Modernity at Large: Cultural Dimensions of Globalization*. Minneapolis: University of Minnesota Press.
Apter, Andrew. 1992. *Black Critics and Kings: The Hermeneutics of Power in Yoruba Society*. Chicago: University of Chicago Press.
Aquina, Sister Mary. 1967. "The people of the spirit—independent church in Rhodesia." *Africa* 37, 203–219.
Arens, William and Ivan Karp. 1989. "Introduction," in *Creativity of Power*, ed. W. Arens and Ivan Karp, xi–xxix. Washington: Smithsonian Institution Press.
Arikpo, Okoi. 1967. *The Development of Modern Nigeria*. Harmondsworth: Penguin Books.
Asad, Talal (ed.). 1973. *Anthropology and the Colonial Encounter*. London: Ithaca Press.
Ashimolowo, Matthew. 1997. *101 Truths about Your Best Friend the Holy Spirit*. London: Mattyson Media.
——— 1999. *Anointing 101*. London: Mattyson Media.
——— 2000a. *101 Answers to Money Problems*. London: Mattyson Media.
——— 2000b. *Breaking Barriers: Tearing down Invisible Barriers: changing Your Speech, Sight, Hearing and Thought from Negative to Positive*. London: Mattyson Media.
Awolalu, J. Omosade. 1970. "The Yoruba philosophy of life." *Présence Africaine* 73, 20–38.
——— 1979. *Yoruba Beliefs and Sacrificial Rites*. Harlow: Longman.
Ayandele, E.A. 1974. *The Educated Elite in the Nigerian Society*. Ibadan: Ibadan University Press.
Baëta, C.G. 1962. *Prophetism in Ghana: A Study of some "Spiritual" Churches*. London: SCM Press.
Barber, Karin. 1991. *I Could Speak until Tomorrow: Oriki, Women and the Past in a Yoruba Town*. Edinburgh: Edinburgh University Press.
——— 2000. *The Generation of Plays: Yoruba Popular Life in Theater*. Bloomington: Indiana University Press.
Barker, Eileen (ed.). 1983. *Of Gods and Man: New Religious Movements in the West*. Macon, GA: Mercer University Press.
——— 1992. *New Religious Movements: A Practical Introduction*. London: HMSO.
Barnes, Sandra T. 1986. *Patrons and Power: Creating a Political Community in Metropolitan Lagos*. Manchester: Manchester University Press.
——— (ed.). 1997a. *Africa's Ogun: Old World and New*. Bloomington: Indiana University Press.
——— 1997b. "The many faces of Ogun," in *Africa's Ogun: Old World and New*, ed. Sandra T. Barnes, 1–26. Bloomington: Indiana University Press.

Barrett, David B. 1968. *Schism and Renewal in Africa: An Analysis of Six Thousand Contemporary Religious Movements*. London: Oxford University Press.

Bascom, William. 1941. "The Sanctions of Ifa Divination." *Journal of the Royal Anthropological Institute* 71:1, 2, 43–45.

———— 1942. "Ifa divination: comments on the paper by J.D. Clarke." *Man* 42:21, 41–45.

———— 1943. "The relationship of Yoruba folklore to divining." *Journal of American Folklore* 56:220, 127–131.

———— 1961. "Odu Ifa: the order of the figures of Ifa." *Bulletin de l'Institut Français d'Afrique Noire* 23:3, 4, 676–682.

———— 1966. "Odu Ifa: the names of the signs." *Africa* 36:4, 408–421.

———— 1969. *Ifa Divination: Communication between Gods and Men in West Africa*. Bloomington: Indiana University Press.

———— 1980. *Sixteen Cowries: Yoruba Divination from Africa to the New World*. Bloomington: Indiana University Press.

Beattie, John. 1966. "Consulting a diviner in Bunyoro. A text." *Ethnology* 5:2, 202–217.

———— 1967a. "Consulting a Nyoro diviner: the ethnologist as client." *Ethnology* 6, 57–65.

———— 1967b. "Divination in Bunyoro, Uganda," in *Magic, Witchcraft, and Curing*, ed. John Middleton, 211–231. New York: Natural History Press.

———— 1969. "Spirit mediumship in Bunyoro," in *Spirit Mediumship and Society in Africa*, ed. John Beattie and John Middleton, 159–170. London: Routledge & Kegan Paul.

Beattie, John and John Middleton (eds.). 1969a. *Spirit Mediumship and Society in Africa*. London: Routledge & Kegan Paul.

———— 1969b. "Introduction," in *Spirit Mediumship and Society in Africa*, ed. John Beattie and John Middleton, xvii–xxx. London: Routledge & Kegan Paul.

Beckman, David M. 1975. *Eden Revival: Spiritual Churches in Ghana*. St. Louis: Concordia Publishing House.

Beier, Ulli. 1956a. "The Egungun cult." *Nigeria Magazine* 51, 380–392.

———— 1956b. "Obatala festival." *Nigeria Magazine* 52, 10–28.

———— 1956c. "Oshun festival." *Nigeria Magazine* 53, 170–187.

———— 1958. "Gelede masks." *Odu* 6, 5–23.

———— 1980. *Yoruba Myths*. Cambridge: Cambridge University Press.

———— 1982. *Yoruba Beaded Crowns: Sacred Regalia of the Olokuku of Okuku*. London: Ethnographica.

Bell, Catherine. 1992. *Ritual Theory, Ritual Practice*. Oxford: Oxford University Press.

Berner, Astrid. 1990. *An Anthropological Study of a North London Aladura Congregation*. Unpublished dissertation, SOAS, University of London.

Berry, Sandra. 1985. *Fathers Work for Their Sons: Accumulation, Mobility, and Class Formation in an Extended Yoruba Community*. Berkeley and Los Angeles: University of California Press.

Bloch, Maurice. 1984. *Marxism and Anthropology*. Oxford: Oxford University Press.

Boddy, Janice. 1989. *Wombs and Alien Spirits: Women, Men, and the Zar Cult in Northern Sudan*. Madison: University of Wisconsin Press.

—— 1994. "Spirit possession revisited: beyond instrumentality." *Annual Review of Anthropology* 23, 407–434.

Booth, Terry. 1984. *We True Christians*. Unpublished dissertation, University of Birmingham.

Bourdieu, Pierre. 1977. *Outline of a Theory of Practice*. London: Cambridge University Press.

Bourguignon, Erika. 1968. "Divination, transe et possession en Afrique transsaharienne," in *La Divination*, vol. II, ed. André Caquot and Marcel Leibovici, 331–358. Paris: Presses Universitaires de France.

—— 1976. "Spirit possession belief and social structure," in *The Realm of the Extra-Human: Ideas and Actions*, ed. Bharati Agehananda, 17–26. The Hague: Mouton Publishers.

Bourguignon, Erika and L. Pettay. 1964. "Spirit possession, trance and cross-cultural research," in *Symposium on New Approaches to the Study of Religion*, ed. June Helm. Seattle: University of Washington Press.

Bourne, Jenny. 1980. "Cheerleaders and ombudsmen: the sociology of race relations in Britain," *Race and Class* 21:4, 331–352.

Bowen, Elenore Smith. 1954. *Return to Laughter*. London: Gollancz.

Brandel-Syrier, Mia. 1962. *Black Woman in Search of God*. London: Lutterworth Press.

Brierley, P. 1999. "Religion in Britain 1900–2000," in *U.K. Christian Handbook*, ed. P. Brierley and H. Wraight, 24–28. London: Christian Research/Harper Collins.

British Council, The. 1969. *Overseas Students in Britain*. London: The British Council.

—— 1970: *Overseas Students in Britain: Statistical Supplement 1969–70*. London: The British Council.

Bruce, Steve. 1996. *Religion in the Modern World: From Cathedrals to Cults*. Oxford: Oxford University Press.

Bruner, Edward M. 1986. "Experience and its expressions," in *The Anthropology of Experience*, ed. Victor Turner and Edward Bruner. Chicago: University of Illinois Press.

Buckley, Anthony. 1985. *Yoruba Medicine*. Oxford: Clarendon Press.

Burridge, Kenelm. 1960. *Mambu: A Melanesian Millennium*. London: Methuen.

—— 1979. *Someone, No One: An Essay on Individuality*. Princeton, NJ: Princeton University Press.

Butler, Eliza M. 1949. *Ritual Magic*. Cambridge: Cambridge University Press.

Calley, Malcolm. 1962. "Pentecostal sects among West Indian migrants," *Race* 3:2, 55–64.

—— 1965. *God's People: West Indian Pentecostal Sects in England*. Oxford: Oxford University Press.

Campbell, Horace. 1980. "Rastafari: culture of resistance." *Race and Class* 22:1, 1–22.

Caquot, André. 1968. "La Divination dans l'ancien Israël," in *La Divination*, vol. I, ed. André Caquot and Marcel Leibovici, 83–113. Parish: Presses Universitaires de France.

Carey, A.T. 1956. *Colonial Students: A Study of the Social Adaption of Coloured Students in London.* London: Secker & Warburg.

Carmichael, Stokely and Chester Hamilton. 1968. *Black Power: The Politics of Liberation in America.* London: Cape.

Carter, Jeffrey. 1997. *The Celestial Church of Christ: Syncretism, Ritual Practice and the Invention of Tradition in a New Religious Movement.* Unpublished dissertation, University of Chicago.

Centre for Contemporary Cultural Studies (CCCS). 1982. *The Empire Strikes Back: Race and Racism in '70s Britain*, ed. CCCS. Birmingham: University of Birmingham Centre for Contemporary Cultural Studies.

Charsley, Simon. 1992. "Dreams in African churches," in *Dreaming, Religion and Society in Africa*, ed. M.C. Jedrej and Rosalind Shaw, 153–176. Leiden: E.J. Brill.

Chevreau, Guy. 1994. *Catch the Fire: The Toronto Blessing.* London: Marshall Pickering.

Clarke, J.D. 1939. "Ifa divination." *Journal of the Royal Anthropological Institute* 69:2, 235–256.

Coleman, James S. 1958. *Nigeria: Background to Nationalism.* Berkeley: University of California Press.

Coleman, Simon. 2000. *The Globalisation of Charismatic Christianity: Spreading the Gospel of Prosperity.* Cambridge: Cambridge University Press.

Comaroff, Jean. 1985. *Body of Power: Spirit of Resistance.* Chicago: University of Chicago Press.

Cornwall, Andrea. 1996. *For Money, Children and Peace: Everyday Struggles in Changing Times in Ado-Odo, Southwestern Nigeria.* Unpublished dissertation, SOAS, University of London.

Corten, André and Ruth Marshall-Fratani. 2001. *Between Babel and Pentecost: Transnational Pentecostalism in Africa and Latin America.* London: C. Hurst & Co.

Cox, Harvey. 1996. *Fire from Heaven.* London: Cassell.

Crapanzano, Vincent. 1977. "Introduction," in *Case Studies in Spirit Possession*, ed. Vincent Crapanzano and Vivian Garrison, 1–40. New York: Witney.

———— 1980. *Tuhami: Portrait of a Moroccan.* Chicago: University of Chicago Press.

Craven, Anna. 1968. *West Africans in London.* London: Institute of Race Relations.

Crowther, Rev. Samuel. 1852. *A Grammar and Vocabulary of the Yoruba Language.* London: Seeleys.

Crumbley, Deidre. 1992. "Impurity and power: women in Aladura churches." *Africa* 62:4, 505–522.

Daley, Patricia. 1996. "Black-Africans: students who stayed," in *Ethnicity in the 1991 Census Vol. II: The Ethnic Minority Population of Great Britain 1966*, ed. Ceri Peach, 44–65. London: HMSO.

Danforth, Loring M. 1989. *Firewalking and Religious Healing: The Anastenaria of Greece and the American Firewalking Movement*. New Jersey: Princeton University Press.

Daniel, W.W. 1968. *Racial Discrimination in England*. Harmondsworth: Penguin.

Davies, Charlotte Aull. 1999. *Reflexive Ethnology: A Guide to Researching Selves and Others*. London: Routledge.

Davis, Ermina. 1977. *In Honour of the Ancestors: the Social Context of Iwi Egungun Chanting in a Yoruba Community*. Unpublished dissertation, Brandein University, Waltham, MA.

Deakin, Nicholas. 1970. *Colour, Citizenship and British Society*. London: Panther Books.

Dennett, R.E. [1910] 1968. *Nigerian Studies, or the Religious and Political System of the Yoruba*. London: Frank Cass.

Deren, Maya. 1953. *The Divine Horsemen: The Voodoo Gods of Haiti*. London: Thames & Hudson.

Desjarlais, Robert. 1992. *Body and Emotion: The Aesthetics of Illness and Healing in the Nepal Himalayas*. Philadelphia: University of Pennsylvania Press.

Devisch, Renaat. 1985. "Perspectives on divination in contemporary sub-Saharan Africa," in *Theoretical Explorations in African Religion*, ed. W. van Binsbergen and M. Schoffeleers. London: Routledge & Keegan Paul.

—— 1991. "Mediumistic divination among the Northern Yaka of Zaire: etiology and ways of knowing," in *African Divination Systems: Ways of Knowing*, ed. Philip M. Peek, 112–132. Bloomington: Indiana University Press.

Drewal, Henry John and M.T. Drewal. [1983] 1990. *Gẹlẹdẹ: Art and Female Power among the Yoruba*. Bloomington: Indiana University Press.

Drewal, Henry John and John Mason. 1998. *Beads, Body and Soul: Art and Light in the Yorùbá Universe*. Los Angeles: UCLA Fowler Museum of Cultural History.

Drewal, Margaret Thompson. 1989. "Dancing for Ògun in Yorubaland and in Brazil," in *Africa's Ogun: Old World and New*, ed. Sandra T. Barnes, 199–234. Bloomington: Indiana University Press.

—— 1992. *Yoruba Ritual: Performers, Play, Agency*. Bloomington: Indiana University Press.

Dumont, Louis. 1985. "A modified view of our origins: the Christian beginnings of modern individualism," in *The Category of the Person*, ed. M. Carrithers, S. Collins, and S. Lukes, 93–122. Cambridge: Cambridge University Press.

Dunham, Katherine. 1994. *Island Possessed*. Chicago: University of Chicago Press.

Durkheim, Emile. [1915] 1964. *Elementary Forms of the Religious Life*. London: George Allen & Unwin.

Eades, J.S. 1980. *The Yoruba Today*. Cambridge: Cambridge University Press.

Eames, Elizabeth. 1992. "Navigating Nigerian bureaucracies; or, 'Why can't you beg?' she demanded," in *The Naked Anthropologist*, ed. Philip DeVita, 184–191. Belmont, CA: Wadsworth Publishing Co.

Echeknube, A.O. 1987. "The question of reincarnation in African religion: a re-appraisal." *Orita* 19:1, 10–26.

Ellis, June. 1971. "Fostering of West African children." *Social Work Today* 2:5, 21–24.

—— (ed.). 1978. *West African Families in Britain. A Meeting of Two Cultures*. London: Routledge & Kegan Paul.

Engels, Frederick. [1884] 1958. *The Origin of the Family, Private Property and the State*. Oxford: Blackwell.

Epega, Afolabi. 1985. *Obi, the Mystical Oracle of Ifa Divination*. New York: Oceanie-Afrique Noire.

Evans-Pritchard, E.E. 1936. Zande Theology. *Sudan Notes and Records* 19 (Khartoum).

—— 1937. *Witchcraft and Oracles among the Azande*. Oxford: Clarendon Press.

—— 1965. *Theories of Primitive Religion*. Oxford: Clarendon Press.

Fadipe, N.A. 1970. *The Sociology of the Yoruba*. Ibadan: University of Ibadan Press.

Fafunwa, A. Babs. 1971. *A History of Nigerian Higher Education*. Lagos: Macmillan & Co. (Nigeria) Ltd.

Fagbola, T.A. 1960. *The Nigerian Student in the U.K. and Western Germany: Focus on His Problems*. Ibadan: Ministry of Information, Western Nigeria.

Farrow, S.S. 1926. *Faith, Fancies and Fetish, or Yoruba Paganism*. London: SPCK.

Fernandez, James W. 1974. "The mission of metaphor in expressive culture." *Current Anthropology* 15:2, 119–145.

—— 1990. "Enclosures: boundary maintenance and the representations over time in Asturian mountain villages (Spain)," in *Culture through Time*, ed. Emiko Ohnuki-Tierney, 94–127. Stanford, CA: Stanford University Press.

—— 1991. "Introduction: confinements of inquiry," in *Beyond Metaphor: The Theory of Tropes in Anthropology*, 1–13. Stanford, CA: Stanford University Press.

Field, M.J. 1960. *Search for Security: An Ethno-Psychiatric Study of Rural Ghana*. London: Faber & Faber.

Fiske, John. 1989. *Television Culture*. London: Routledge.

Fogelson, Raymond D. and Richard N. Adams (eds.). 1977. *The Anthropology of Power*. London: The Academic Press.

Foot, Paul. 1965. *Immigration and Race in British Politics*. Harmondsworth: Penguin Books.

—— 1969. *The Rise of Enoch Powell*. Harmondsworth: Penguin Books.

Foucault, Michel. 1972. *The Archaeology of Knowledge*. London: Tavistock.

—— 1980. *Power and Knowledge*. ed. Colin Gordon. Harlow: Pearson.

Freston, Paul. 2001. "The transnationalisation of Brazilian Pentecostalism: the Universal Church of the Kingdom of God," in *Between Babel and Pentecost: Transnational Pentecostalism in Africa and Latin America*, ed. André Corten and Ruth Marshall-Fratani, 196–215. London: C. Hurst & Co.

Fryer, Peter. 1984. *Staying Power: The History of Black People in Britain*. London: Pluto.

Gbadegesin, Olusegun. 1991. "Negritude and its contribution to the civilization of the universal: Leopold Senghor and the question of ultimate reality and meaning." *Ultimate Reality and Meaning* 14:1, 30–45.

Geertz, Clifford. 1973. *The Interpretation of Cultures*. New York: Basic Books.

———— 1983. *Local Knowledge: Further Essays in Interpretative Anthropology*. New York: Basic Books.

Gerloff, Roswith. 1992. *A Plea for British Black Theologies: The Black Church Movement in Britain in Its Transatlantic Cultural and Theological Interaction* (2 vols.). New York: Peter Lang.

———— 1999. "The significance of the African Christian diaspora to Europe: a report on four meetings in 1997/8." *Journal of Religion in Africa* 29:1.

Giddens, Anthony. 1976. *New Rules of Sociological Method*. London: Hutchinson & Co.

———— 1989. *Sociology*. Cambridge: Polity Press.

Gifford, Paul. 1998. *African Christianity and Its Public Role*. London: C. Hurst & Co.

———— 2001. "The complex provenance of some elements of African Pentecostal theology," in *Between Babel and Pentecost: Transnational Pentecostalism in Africa and Latin America*, ed. André Corten and Ruth Marshall-Fratani, 62–79. London: C. Hurst & Co.

Gilroy, Paul. 1987. "*There Ain't No Black in the Union Jack*": *The Cultural Politics of Race and Nation*. London: Routledge.

Gleason, Judith. 1973. *A Recitation of Ifa, Oracle of the Yoruba*. New York: Grossman Publishers.

———— 1987. *Oya: in Praise of the Goddess*. Boston: Shambhala.

Goodman, Felicitas D. 1972. *Speaking in Tongues: A Cross-Cultural Study of Glossolalia*. Chicago: University of Chicago Press.

———— 1988. *How about Demons? Possession and Exorcism in the Modern World*. Bloomington: Indiana University Press.

Goody, Esther. 1971. "The varieties of fostering." *New Society* 18:462, 237–239.

Goody, Esther and Christine Muir. 1972. *Factors Related to the Delegation of Parental Roles among West Africans in London*. Unpublished report, University of Cambridge: Committee for the Social and Political Sciences.

Goody, Esther and Christine Muir Groothues. 1977. "The West Africans: the quest for education," in *Between Two Cultures: Migrants and Minorities in Britain*, ed. James Watson, 151–180. Oxford: Basil Blackwell.

———— 1979. "Stress in marriage: West African Couples in London ," in *Minority Families: Support and Stress*, ed. Verity Saifullah Khan, 59–86. London: Macmillan Press.

Goody, Jack. 1993. *The Culture of Flowers*. Cambridge: Cambridge University Press.

Great Britain. 1964. Home Office. *Commonwealth Immigrants Act 1962: Control of Immigration Statistics, 1962/3*. London: H.M.S.O.

—— 1967. General Register Office. *Sample Census 1966*. London: H.M.S.O.

—— 1968. Home Office. *Commonwealth Immigrants Act 1962: Control of Immigration Statistics, 1967*. London: H.M.S.O.

—— 1972. *Matrimonial Proceedings (Polygamous Marriages) Act*. London: H.M.S.O.

—— 1974. Office of Population Censuses and Surveys. *Census 1971*. London: H.M.S.O.

Hackett, Rosalind. 1983. "Power and authority in Nigerian independent churches." *West African Religion* 20:1, 2, 37–54.

—— 1987a. *New Religious Movements in Nigeria*. Lewiston, NY: Edwin Mellen Press.

—— 1987b. "Introduction: variation on a theme," in *New Religious Movements in Nigeria*, ed. Rosalind Hackett, 1–17. Lewiston, NY: Edwin Mellen Press.

—— 1987c. "Thirty years of growth and change in a West African independent church: a social perspective," in *New Religious Movements in Nigeria*, ed. Rosalind Hackett, 161–177. Lewiston, NY: Edwin Mellen Press.

—— 1987d. "Women as leaders and participants in the spiritual churches," in *New Religious Movements in Nigeria*, ed. Rosalind Hackett, 191–208. Lewiston, NY: Edwin Mellen Press.

—— 1991. "New religious movements," in *Religion and Society in Nigeria: Historical and Sociological Perspectives*, ed. Jacob Kehinde and Toyin Falola, 282–300. Ibadan: Spectrum Books.

—— 1993. "The symbolics of power discourse among contemporary religious groups in West Africa," in *Religious Transformations and Socio-Political Change*, ed. Luther Martin, 381–409. Berlin: Mouton de Gruyter

—— 1995. "The gospel of prosperity in West Africa," in *Religion and the Transformation of Capitalism*, ed. R. Roberts, 199–214. London: Routledge.

—— 1998. "Charismatic/pentecostal appropriation of media technologies in Nigeria and Ghana." *Journal of Religion in Africa* 18:3, 258–277.

Hagan, George P. 1988. "Divinity and experience: the trance and Christianity in Southern Ghana," in *Vernacular Christianity: Essays in the Social Anthropology of Religion Presented to Godfrey Lienhardt*, ed. Wendy James and Douglas Johnson, 146–156. Oxford: JASO.

Hall, Stuart. 1978. "Racism and reaction," in *Five Views of Multi-Racial Britain*, ed. Commission for Racial Equality. London: CRE.

Hallen, Barry and J.O. Sodipo. 1986.*Knowledge, Belief and Witchcraft: Analytic Experiments in African Philosophy*. London: Ethnographica.

Hammond, Phillip (ed.). 1985. *The Sacred in a Secular Age*. Berkeley: University of California Press.

Harper, Michael. 1968. *Walk in the Spirit*. London: Hodder and Stoughton.

Harris, Hermione. 1972. "Black women and work," in *The Body Politic: Women's Liberation in Britain 1969–1972*, ed. M. Wandor, 166–174. London: Stage I.

Harrison, Simon. 1992. "Ritual as intellectual property." *Man* 27:2, 225–244.

Hastrup, Kirsten and Peter Hervik (eds.). 1994. *Social Experience and Anthropological Knowledge*. London: Routledge.

Hervik, Peter. 1994. "Shared reasoning in the field," in *Social Experience and Anthropological Knowledge*, ed. Kirsten Hastrup and Peter Hervik, 78–100. London: Routledge.

Hill, Clifford S., 1963. *West Indian Migrants and the London Churches*. London: Oxford University Press.

Hollenweger, W.J. 1972. *The Pentecostals*. London: SCM Press.

Hollis, Martin, and Steven Lukes. 1982. *Rationality and Relativism*. Oxford: Blackwell.

Horton, Robin. 1967. "African traditional thought and Western science. Pt. 2: The 'closed' and 'open' predicaments." *Africa* 37:2, 155–187.

——— 1971. "African conversion." *Africa* 41:2, 85–108.

Howard, Rowland. 1996. *The Rise and Fall of the Nine o'Clock Service*. London: Mobray.

Hunt, Geoffrey P. 1973. *Transmission of Knowledge within a West African Immigrant Religious Group*. Unpublished dissertation, University of London.

Hunt, Stephen and Nicola Lightly. 2001. "The British black Pentecostal 'revival': identity and belief in the 'new' Nigerian churches." *Ethnic and Racial Studies* 24:1, 104–124.

Idowu, E. Bolaji. 1962. *Olódùmarè: God in Yoruba Belief*. London: Longman.

——— 1970. "The challenge of witchcraft." *Orita* 4:1, 3–16.

Ilesanmi, Thomas M. 1983. "On power among the Yoruba: a literary perspective." *Odu* 23, 103–115.

——— 1993. "Language of African traditional religions." *Research in Yoruba Language and Literature* 4, 63–67.

Inglis, Brian. 1990. *Trance: A Natural History of Altered States of Mind*. London: Paladin Books.

Irele, Abiola. 1975. "Tradition and the Yoruba writer: D.O. Fagunwa, Amos Tutuola and Wole Soyinka." *Odu* 11, 75–100.

James, William. [1902] 1929. *The Varieties of Religious Experience*. New York: Random House.

Jenkins, Robin. 1971. *The Production of Knowledge in the IRR*. London: Independent Labour Party.

Jilek, Wolfgang G. 1989. "Therapeutic use of altered states of consciousness in contemporary North American Indian dance ceremonials," in *Altered States of Consciousness and Mental Health: A Cross-Cultural Perspective*, ed. Colleen Ward, 167–185. London: Sage.

Johnson, Samuel. [1921] 1969. *The History of the Yorubas*. London: Routledge & Kegan Paul.

Jules-Rosette, Bennetta. 1975. *African Apostles*. Ithaca, NY: Cornell University Press.

—— 1978. "The veil of objectivity: prophecy, divination and social enquiry." *American Anthropologist* 80:3, 549–570.

—— 1979. *The New Religions of Africa*. Norwood, NJ: Ablex Publishing Corp.

Kalilombe, P. 1997. "Black Christianity in Britain." *Ethnic and Racial Studies* 20:2, 306–324.

Kaser, Richard T. 1995. *African Oracles in 10 Minutes*. New York: Avon Books.

Keesing, R. 1984. "Rethinking *Mana*." *Journal of Anthropological Research* 40, 137–156.

Kendall, Michael. 1968. *Overseas Students in Britain—An Annotated Bibliography*. Research Unit for Student Problems (University of London) with United Kingdom Council for Overseas Student Affairs.

Kerridge, Roy. 1995. *The Storm Is Passing Over: A Look at Black Churches in Britain*. London: Thames and Hudson.

Kiernan, J.P. 1990. *The Production and Management of Therapeutic Power in Zionist Churches within a Zulu City*. Lewiston: The Edwin Mellen Press.

Killingray, David (ed.). 1994a. *Africans in Britain*. Ilford: Frank Cass.

—— 1994b. "Africans in the United Kingdom: an introduction," in *Africans in Britain*, ed. David Killingray, 2–27. Ilford: Frank Cass.

Kingston, Charles. [1939] 1965. *Fullness of Power: Talks on the Gifts of the Holy Spirit*. London: Elim Publishing House.

Kramer, Fritz W. 1993. *The Red Fez: Art and Spirit Possession in Africa*. London: Verso.

Kulick, Don, and Margaret Willson. 1995. *Taboo: Sex, Identity and Erotic Subjectivity in Anthropological Fieldwork*. London: Routledge.

La Barre, Weston. 1972. *The Ghost Dance: The Origins of Religion*. New York: Dell Publishing Co.

La Fontaine, Jean S. 1985. "Person and individual: some anthropological reflections," in *The Category of the Person*, ed. Michael Carrithers, Steven Collings, and Steven Lukes, 123–140. Cambridge: Cambridge University Press.

Lambek, Michael. 1981. *Human Spirits: A Cultural Account of Trance in Mayotte*. Cambridge: Cambridge University Press.

—— 1989. "From disease to discourse: remarks on the conceptualisation of trance and spirit possession," in *Altered States of Consciousness and Mental Health: A Cross-Cultural Perspective*, ed. Colleen Ward, 36–61. London: Sage.

Lambo, T. Adeoye. 1960. "Characteristic features of the psychology of the Nigerian." *West African Medical Journal* 9:3, 95–104.

—— 1963. *African Traditional Beliefs: Concepts of Health and Medical Practice*. Ibadan: Ibadan University Press.

—— 1964. "Patterns of psychiatric care in developing African countries," in *Magic, Faith and Healing: Studies in Primitive Psychiatry Today*, ed. Ari Kiev, 443–453. Glencoe: The Free Press.

Lawrence, Peter. 1964. *Road Belong Cargo: A Study of the Cargo Movement in the Southern Madang District, New Guinea.* Manchester: Manchester University Press.

Lawuyi, Olatunde. 1986. "No king as God: towards an understanding of Yoruba slogans." *Odu* 29, 102–114.

——— 1988. "The world of the Yoruba taxi driver: an interpretative approach to vehicle slogans." *Africa* 58:1, 1–13.

——— 1991. "Self-potential as a Yoruba ultimate." *Ultimate Reality and Meaning* 14:1, 21–29.

Lee, Raymond L.M. 1989. "Self-presentation in Malaysian spirit seances: a dramaturgical perspective on altered states of consciousness," in *Altered States of Consciousness and Mental Health: a Cross-Cultural Perspective,* ed. Colleen Ward, 251–266. London: Sage.

Lee, S.G. 1969. "Spirit Possession among the Zulu," in *Spirit Mediumship and Society in Africa,* ed. John Beattie and John Middleton, 128–156. London: Routledge & Kegan Paul.

Legesse, Asmarom. 1994. "Prophetism, democharisma, and social change," in *Religion in Africa: Experience and Expression,* ed. Thomas Blakely, Walter van Beek, and Dennis Thomson, 314–341. London: James Currey.

Leighton, Alexander, T, Adeoye Lambo, et al. 1963. *Psychiatric Disorder among the Yoruba.* Ithaca: Cornell University Press.

Levine, Robert, Nancy Klein, and Constance Owen. 1967. "Father–child relationships and changing life-styles in Ibadan, Nigeria," in *The City in Modern Africa,* ed. Horace Miner, 215–255. London: Pall Mall Press.

Lewis, Ioan. 1966. "Spirit possession and deprivation cults." *Man* 1:3, 307–329.

——— 1967. "Spirits and sex war." *Man* 2:4, 626–628.

——— 1971. *Ecstatic Religion: An Anthropological Study of Spirit Possession and Shamanism.* Harmondsworth: Penguin Books.

——— 1986. *Religion in Context: Cults and Charisma.* Cambridge: Cambridge University Press.

Lillie, D.G. 1966. *Tongues under Fire.* London: The Fountain Trust.

Little, Kenneth and Anne Price. 1967. "Some trends in modern marriage among West Africans." *Africa* 37, 407–423.

Lloyd, Barbara. 1970. "Yoruba mothers' reports of childrearing: some theoretical and methodological considerations," in *Socialization: the Approach from Social Anthropology,* ed. P. Mayer, 75–107. London: Tavistock Publications.

Lloyd, Peter.1966a. "Introduction," in *The New Elites of Tropical Africa,* ed. Peter Lloyd, 1–65. Oxford: Oxford University Press.

——— 1966b. "Class Consciousness among the Yoruba," in *The New Elites of Tropical Africa,* ed. Peter Lloyd, 328–340. London: Oxford University Press.

——— 1967. "The Elite," in *The City of Ibadan,* ed. Peter C. Lloyd, A.L. Mabogunge, and B. Awe, 129–150. Cambridge: Cambridge University Press.

——— 1974. *Power and Independence: Urban Africans' Perception of Social Inequality.* London: Routledge & Kegan Paul.

Lucas, J. Olumide. 1948. *The Religion of the Yoruba.* Lagos: CMS.

Ludwar-Ene, Gundrun. 1991. "Spiritual church participation as a survival strategy among urban migrant women in Southern Nigeria," in *New Religious Movements and Society in Nigeria*, ed. Gundrun Ludwar-Ene, 53–69. Bayreuth: Bayreuth University.

Luhrmann, T.M. 1992. *Persuasions of a Witch's Craft: Ritual Magic in Contemporary England.* Oxford: Blackwell.

Lukes, Steven. 1973. *Individualism.* Oxford: Blackwell.

Lutz, Catherine A. and Lila Abu-Lughod (eds.). 1990. *Language and the Politics of Emotion.* Cambridge: Cambridge University Press.

Macklin, June. 1977. "A Connecticut Yankee in Summer Land," in *Case Studies in Spirit Possession*, ed. Vincent Crapanzano and Vivian Garrison, 41–85. New York: Wiley.

Maclean, Ulna. 1971. *Magical Medicine, a Nigerian Case Study.* London: Allen Lane.

Macpherson, William. 1999. *The Stephen Lawrence Inquiry.* London: The Stationary Office.

Maier, Karl. 2000. *This House Has Fallen: Nigeria in Crisis.* Harmondsworth: Allen Lane.

Makinde, M. Akin. 1983. "Ifa as a repository of knowledge." *Odu* 23, 116–121.

Malinowski, B. 1954. "Magic, science and religion," in *Magic, Science and Religion and Other Essays*, 17–92. New York: Doubleday Anchor Books.

——— 1967. *A Diary in the Strict Sense of the Term.* New York: Harcourt, Brace and World. Inc.

Marett, R. 1932. *Faith, Hope and Charity in Primitive Religion.* Oxford: Oxford University Press.

Marris, Peter. 1961. *Family and Social Change in an African City.* London: Routledge & Kegan Paul.

Marshall, Ruth. 1993. "Power in the name of Jesus: social transformation and Pentecostalism in Western Nigeria 'revisited,' " in *Legitimacy and the State in Twentieth Century Africa*, ed. Terence Ranger and Olufemi Vaughan, 213–246. Oxford: Macmillan.

——— 1995. " 'God is not a democrat': Pentecostalism and democratisation in Nigeria," in *The Christian Churches and the Democratisation of Africa*, ed. Paul Gifford, 239–260. Leiden: Brill.

Marshall-Fratani, Ruth. 1998. "Mediating the global and local in Nigerian Pentecostalism." *Journal of Religion in Africa* 28:3, 278–315.

Martin, David. 1990. *Tongues of Fire: The Explosion of Protestantism in Latin America.* Oxford: Blackwell.

——— 2002. *Pentecostalism: The World Their Parish.* Oxford: Blackwell.

Marx, Karl and Frederick Engels. 1957. *On Religion.* Moscow: Progress Publishers.

Matory, J. Lorand. 1994. *Sex and the Empire that is no more: Gender and the Politics of Metaphor in Oyo Yoruba Religion.* Minneapolis: University of Minnesota Press.

Mayo, Marjorie. 1969. "West African voluntary associations." *Community Development Journal* 4:4, 212–215.

Mbon, Friday M. 1987. "Public response to new religious movements in contemporary Nigeria," in *New Religious Movements in Nigeria,* ed. Rosalind Hackett, 209–235. Lewiston, NY: The Edwin Mellen Press.

———— 1991. "The Social Impact of Nigeria's New Religious Movements," in *New Religious Movements and Rapid Social Change,* ed. James Beckford, 177–196. London: Sage.

———— 1992. *Brotherhood of the Cross and Star: A New Religious Movement in Nigeria.* Frankfurt-am-Main: Lang.

McClelland, E.M. 1966. "The significance of number in the Odu of Ifa." *Africa* 36:4, 421–430.

McKenzie, Peter. 1976. *Inter-Religious Encounters in West Africa: Samuel Ajayi Crowther's Attitude to African Traditional Religion and Islam.* Leicester: University of Leicester.

———— 1992. "Dreams and Visions in Nineteenth Century Yoruba Religion," in *Dreaming, Religion and Society in Africa,* ed. M.C. Jedrej and Rosalind Shaw, 126–134. Leiden: Brill.

Meek, C.K. 1943. *The Religions of Nigeria.* Africa 14:3, 106–118.

Mendonsa, E.L. 1982. *The Politics of Divination: A Processual View of Reactions to Illness and Deviance among the Sisala of Ghana.* Berkeley: University of California Press.

Mernissi, Fatima. 1995. *The Harem Within: Tales of a Moroccan Girlhood.* London: Bantam Books.

Meyer, Piet. 1991. "Divination among the Lobi of Burkina Faso," in *African Divination Systems: Ways of Knowing,* ed. Philip Peek, 91–100. Bloomington: Indiana University Press.

Middleton, John. 1969. "Oracles and Divination among the Lugbara," in *Man in Africa,* ed. Mary Douglas and Phyllis Kaberry, 261–277. London: Tavistock Publications.

Middleton, John and E.H. Winter (eds.). 1963. *Witchcraft and Sorcery in East Africa.* London: Routledge & Kegan Paul.

Mitchell, John. 1997. "A moment with Christ: the importance of feelings in the analysis of belief." *Journal of the Royal Anthropological Institute* 3, 79–94.

Mitchell, R.C. 1963. "Christian Healing," in *African Independent Church Movements,* ed. Victor Hayward, 47–51. London: Edinburgh House Press.

Moller, Caroline. 1968. *An Aladura Church in Eastern Nigeria.* Unpublished dissertation, SOAS, University of London.

Moore, Omar Khayyam. 1957. "Divination—a new perspective." *American Anthropologist* 59, 69–74.

Moore, Robert. 1975. *Racism and Black Resistance in Britain.* London: Pluto Press.

Morris, B.S. and I.O. Ajijola. 1967. *International Community?* London: National Union of Students.

Morris, Brian. 1994. *Anthropology of the Self: The Individual in Cultural Perspective.* London: Pluto Press.

Morton-Williams, Peter. 1956a. "The Atinga cult among the South Western Yoruba: a sociological analysis of a witch-finding movement." *Bulletin de l'IFAN* 28, séries B:18 (3–4), 315–334.

—— 1956b. "The Egungun society," in *Proceedings of the 3rd Annual Conference of the West African Institute of Social and Economic Research,* University College, Ibadan, 90–102.

—— 1960a. "Yoruba responses to the fear of death." *Africa* 30:1, 34–40.

—— 1960b. "The Yoruba Ogboni cult in Oyo." *Africa* 30:4, 362–374.

—— 1964. "An outline of the cosmology and cult organization of the Oyo Yoruba." *Africa* 34:3, 243–261.

—— 1966. "The mode of (Ifa) divination." *Africa* 36:4, 406–408.

—— 1967. *Processes of Change in the Social Organisation of some Yoruba Tribes in South-West Nigeria.* Unpublished dissertation, University of London.

Muir, Christine and Esther Goody. 1970. *Preliminary Report of a Survey of West African Families.* Unpublished report for the Social Science Research Council.

—— 1972. "Student parents: West African families in London." *Race* 13:3, 329–336.

Nassau, R.H. 1904. *Fetichism in West Africa.* London: Duckworth & Co.

Nelson, Geoffrey. 1969. *Spiritualism and Society.* London: Routledge & Kegan Paul.

Nigeria: Federal Ministry of Economic Development, National Manpower Board. 1965. *A Study of Nigerian Professional Manpower in Selected Occupations.* Manpower Study No. 3. Lagos.

Nigeria: Federal Ministry of Economic Development, National Manpower Board. 1967. *Nigeria's Professional Manpower in Selected Occupations.* Manpower Study No. 5. Lagos.

Nigeria: Federal Ministry of Education. 1960. *Investment in Education.* The Report of the Commission on Post-School Certificate and Higher Education in Nigeria (the Ashby Report). Lagos.

Nigeria: Federal Ministry of Information. 1968. *The Policy of the Federal Military Government on Statutory Corporations and State-owned Companies.* Lagos.

Nigeria Magazine. 1957. "Cherubim and Seraphim." 53, 119–134.

Oesterreich, T.K. 1930. *Possession, Demoniacal and Other, among Primitive Races, in Antiquity, the Middle Ages and in Modern Times.* London: Kegan Paul & Co.

Offiong, E. 1987. "Schism and religious independency in Nigeria: the case of the Brotherhood of the Cross and Star," in *New Religious Movements in Nigeria,* ed. Rosalind Hackett, 179–191. Lewiston, NY: Edwin Mellen Press.

Oguntomilade, Jacob. 1987. *Father E. Olu Coker: A Charismatic Star of the Cherubims.* Lagos: Landmark Publications.

Oguntoye, Jide. 1987. *Come Home My Love*. Ibadan: Paperback Publishers Ltd.

Ojo, Matthews A. 1987. *The Growth of Campus Christianity and Charismatic Movements in Western Nigeria*. Unpublished dissertation, Kings College, University of London.

——— 1988. "The contextual significance of the charismatic movements in Independent Nigeria." *Africa* 58:2, 175–192.

——— 1995. "The charismatic movement in Nigeria today." *International Bulletin of Missionary Research* 19:3, 114–118.

——— 1997. "Communality and empowerment: *African Charismatics in London*." Unpublished paper, consultation on the significance of the African Religious Diaspora in Europe, University of Leeds.

Okafor, Nduka. 1971. *The Development of Universities in Nigeria*. London: Longman.

Okediji, Francis Olu. 1967. "Some social psychological aspects of fertility among married women in an African city." *Nigerian Journal of Economic and Social Studies* 9:1, 67–79.

Okely, Judith. 1994. "Vicarious and sensory knowledge of chronology and change: ageing in rural France," in *Social Experience and Anthropological Knowledge*, ed. Kirsten Hastrup and Peter Hervick, 45–64. London: Routledge.

——— 1996a. "The self and scientism," in *Own or Other Culture*, ed. Judith Okely, 27–44. London: Routledge.

——— 1996b. "Fortune-tellers: fakes or therapists?" in *Own or Other Culture*, ed. Judith Okely, 94–114. London: Routledge.

Okely, Judith and Helen Callaway (eds.). 1992. *Anthropology and Autobiography*. London: Routledge.

Olabimtan, Afolabi. 1975. "Religion as a theme in Fagunwa's novels." *Odu* 11, 101–114.

Olayiwola, David O. 1987. "The Aladura: its strategies for mission and conversion in Yorubaland, Nigeria." *Orita* 19:1, 40–56.

Olukoga, Funso. 2002. "Attracting investments from Nigerians abroad." *Makela Magazine* 1, 15–17.

Olupona, Jacob Kehinde. 1987. "The Celestial Church of Christ in Ondo: a phenomenological perspective," in *New Religious Movements in Nigeria*, ed. Rosalind Hackett, 45–73. Lewiston, NY: Edwin Mellen Press.

Olupona, Jacob Kehinde and Toyin Falola (eds.). 1991. *Religion and Society in Nigeria: Historical and Sociological Perspectives*. Ibadan: Spectrum Books.

Olusanya, P.O. 1967. "The educational factor in human fertility: a case study of the residents of a suburban area in Ibadan, Western Nigeria." *Nigerian Journal of Economic and Social Studies* 9:3, 351–374.

Omotoso, Kole. 1971. *The Edifice*. London: Heinemann.

Omoyajowo, J. Akinyele. 1978. "The Aladura churches in Nigeria since Independence," in *Christianity in Independent Africa*, ed. Edward Fasholé-Luke et al., 96–110. London: Rex Collings.

—— 1982. *Cherubim and Seraphim: The History of an African Independent Church*. New York: Nok Publishers International Ltd.

—— 1984. *Diversity in Unity: The Development and Expansion of the Cherubim and Seraphim Church of Nigeria*. Lanham: University Press of America Inc.

Oro, Ari Pedro and Pablo Semán. "Brazilian Pentecostalism crosses national borders," in *Between Babel and Pentecost*, ed. André Corten and Ruth Marshall-Fratani, 181–195. London: C. Hurst & Co.

Osoba, Segun. 1977. "The Nigerian power élite, 1952–65," in *African Social Studies*, ed. C.W. Gutkind and P. Waterman, 368–382. London: Heinemann.

Osofisan, 'Sola. 1991. *The Living and the Dead*. Nigeria: Heinemann Educational Books Plc.

Osunwole, Samuel. 1991. "Witchcraft and sorcery: Yoruba beliefs and medicine." *Orita* 23:2, 73–82.

Owusu, Kwesi (ed.). 2000. *Black British Culture and Society*. London: Routledge.

Oyedepo, David O. 1993. *Releasing the Supernatural: An Adventure into the Spirit World*. Lagos: Dominion Publishing House.

—— 1996. *The Release of Power*. Lagos: Dominion Publishing House.

Oyegoke, Lekan. 1984. *Laughing Shadows*. Harlow, Essex: Longman.

Oyetade, B. Akintunde. ca. 1996. *Ọta: Enemy in Yoruba Belief*. Unpublished article.

Oyewumi, Oyeronke. 1997. *The Invention of Women: Making an African Sense of Western Gender Discourses*. Minneapolis: University of Minnesota Press.

Palmié, Stephen. 1995. "Against syncretism: 'Africanizing' and 'Cubanizing' discourses in North America òrìsà worship," in *Counterworks: Managing the Diversity of Knowledge*, ed. Richard Fardon, 73–104. London: Routledge.

Park, George K. 1963. "Divination and its social contexts." *Journal of the Royal Anthropological Institute* 93:2, 195–209.

Parkin, David. 1991. "Simultaneity and sequencing in the oracular speech of Kenyan diviners," in *African Divination Systems: Ways of Knowing*, ed. Philip Peek, 173–189. Bloomington: Indiana University Press.

Parrinder, E.G. 1951. *West African Psychology*. London: Lutterworth.

—— [1949] 1961. *West African Religion*. London: Epworth Press.

—— 1953. *Religion in an African City*. Oxford: Oxford University Press.

Peek, Philip (ed.). 1991a. *African Divination Systems: Ways of Knowing*. Bloomington: Indiana University Press.

—— 1991b. "The study of divination, present and past," in *African Divination Systems: Ways of Knowing*, ed. Philip Peek, 1–22. Bloomington: Indiana University Press.

—— 1991c. "African divination systems: non-normal modes of cognition," in *African Divination Systems: Ways of Knowing*, ed. Philip Peek, 193–212. Bloomington: Indiana University Press.

Peel, J.D.Y. 1967. "Religious change among the Yoruba." *Africa* 37, 292–306.

—— 1968. *Aladura: A Religious Movement among the Yoruba*. Oxford: Oxford University Press.

—— 1978. "*Olaju*: A Yoruba concept of development." *Journal of Development Studies* 14, 135–165.

—— 1983. *Ijeshas and Nigerians*. Cambridge: Cambridge University Press.

—— 1990. "The pastor and the *babalawo*: the encounter of religions in nineteenth-century Yorubaland." *Africa* 60:3, 338–369.

—— 1993. "Between Crowther and Ajayi: the religious origins of the Nigerian intelligentsia," in *African Historiography: Essays in Honour of Jacob Ade Ajayi*, ed. Toyin Falola, 64–79. Harlow: Longman.

—— 1994. "Historicity and pluralism in some recent studies of Yoruba religion." *Africa* 64:1, 150–166.

—— 2000. *Religious Encounter and the Making of the Yoruba*. Bloomington: Indiana University Press.

—— 2002. "Gender in Yoruba religious change." *Journal of Religion in Africa* 32, 1–31.

Platt, William. 1935. *From Fetish to Faith: The Growth of the Church in West Africa*. London: Edinburgh House Press.

Plotnicov, Leonard. 1967. *Strangers to the City: Urban Man in Jos, Nigeria*. Pittsburgh: University of Pittsburgh Press.

—— 1970. "The composition and role of the modern African elite in a middle-sized Nigerian city," in *Social Stratification in Africa*, ed. A. Tuden and L. Plotnicov, 269–302. New York: Free Press.

Poewe, Karla (ed.). 1994. *Charismatic Christianity as a Global Culture*. Columbia: University of South Carolina Press.

Post, Kenneth. 1968. *The New States of West Africa*. Harmondsworth: Penguin.

Pressel, Esther. 1974. "Umbanda trance and possession in São Paulo, Brazil," in *Trance, Healing, and Hallucination; Three Field Studies in Religious Experience*, ed. Felicitas Goodman, Jeannette Henney, and Esther Pressel, 113–225. New York: John Wiley & Sons.

Prince, Raymond. 1961. "The Yoruba image of the witch." *Journal of Mental Science* 107, 795–805.

—— 1964. "Indigenous Yoruba psychiatry," in *Magic, Faith and Healing*, ed. Ari Kiev, 84–120. Glencoe: Free Press.

Probst, Peter. 1989. "The letter and the spirit: literacy and religious authority in the history of the Aladura Movement in Western Nigeria." *Africa* 59:4, 478–495.

Pro Mundi Vita. 1989. *Does Development Lead to Secularisation?* Study no. 11. Belgium: R.C. Information and Research Centre.

Rabinow, Paul. 1977. *Reflections on Fieldwork in Morocco*. Berkeley: University of California Press.

Rack, Philip. 1982. *Race, Culture, and Mental Disorder*. London: Tavistock Publications.

Ray, Benjamin. 1993. "Aladura Christianity: a Yoruba religion." *Journal of Religion in Africa*, 23:3, 266–291.

Retel-Laurentin, Anne. 1974. "La Force de la Parole," in *Divination et Rationalité*, ed. J.P. Vernant et al., 295–319. Paris: Seuil.

Rex, John and David Mason (eds.). 1986. *Theories of Race and Ethnic Relations*. Cambridge: Cambridge University Press.

Richardson, James. 1985. "Studies of conversion: secularization or re-enchantment?" in *The Sacred in a Secular Age*, ed. Philip E. Hammond, 104–121. Berkeley: University of California Press.

Robbins, Anthony. 1988. *Unlimited Power: The New Science of Personal Achievement*. London: Simon and Schuster.

Robbins, Thomas. 1986. *Cults, Converts and Charisma: The Sociology of New Religious Movements*. London: Sage Publications.

Robertson, R. 1992. *Globalisation: Social Theory and Global Culture*. London: Sage.

Rose, E.J.B. 1969. *Colour and Citizenship*. Oxford: Institute of Race Relations.

Rouse, Roger. 1995. "Questions of identity: personhood and collectivity in transnational migration to the United States." *Critique of Anthropology* 15:4, 351–380.

Rowland, J. 1973. *Community Decay*. Harmondsworth: Penguin Books.

Sebald, H. 1988. "The 6th and 7th Books of Moses: the historical and sociological vagaries of a grimoire." *Ethnologia Europaea* 18:1, 53–58.

Seligman, C.G. 1932. *Pagan Tribes of the Nilotic Sudan*. London: G. Routledge.

Shaw, Rosalind. 1991. "Splitting truths from darkness: epistemological aspects of Temne divination," in *African Divination Systems: Ways of Knowing*, ed. Philip Peek, 137–152. Bloomington: Indiana University Press.

Shelton, A.J. 1965. "The meaning and method of Afa divination among the Northern Nsukka Ibo." *American Anthropologist* 67, 1441–1454.

Simpson, George. 1956. "Jamaican revivalist cults." *Social and Economic Studies* 5:4, 321–442.

Sivanandan, A. 1969. *Coloured Immigrants in Britain: A Select Bibliography*. London: Institute of Race Relations.

——— 1976. "Race, class and the state: the Black experience in Britain." *Race and Class* 17:4, 347–368.

——— 1983. "Challenging racism: strategies for the 80s." *Race and Class* 25:2, 1–11.

The Sixth and Seventh Books of Moses. n.d. Chicago: De Laurence.

Skultans, Vieda. 1971. "The healing process." *New Society* 17:454, 992–994.

Smith, D.J. 1977. *Racial Disadvantage in Britain*. Harmondsworth: Penguin Books.

Smythe, Hugh H. and Mabel M. Smythe. 1960. *The Nigerian Elite*. California: Stanford University Press.

Soyinka, Wole. [1981] 1983. *Ake: The Years of Childhood*. London: Arena.

St. Aubin, Lorna. 1990. *The New Age in a Nutshell: A Guide to Living in New Times*. Bath: Gateway Books.

Stapleton, Pat. 1978. "The West African background," in *West African Families in Britain: A Meeting of Two Cultures*, ed. June Ellis, 14–38. London: Routledge & Kegan Paul.

Stevens, Philip. 1965. "The Festival of the Images at Esie." *Nigeria Magazine* 87, 237.

Stocking, George W. (ed.). 1991. *Colonial Situations: Essays on the Contextualization of Ethnographic Knowledge*. Madison, WI: University of Wisconsin Press.

Stoller, Paul. 1989a. "Stressing social change and Songhay possession," in *Altered States of Consciousness and Mental Health*, ed. Colleen Ward, 267–284. London: Sage.

———— 1989b. *The Taste of Ethnographic Things: The Senses in Anthropology*. Philadelphia: University of Pennsylvania Press.

———— 1995. *Embodying Colonial Memories: Spirit Possession, Power, and the Hauka in West Africa*. London: Routledge.

Stoller, Paul and Cheryl Olkes. 1987. *In Sorcery's Shadow*. Chicago: University of Chicago Press.

Stone, R.H. 1900. *In Africa's Forest and Jungle, or 6 Years among the Yorubans*. London: Oliphant Anderson and Ferrier.

———— 1965. *Yoruba Lore and the Universe*. Occasional paper no. 4. Ibadan: Institute of Education, University of Ibadan.

———— 1969. *Yoruba Concepts of the Natural World in Relation to Learning Science*. Unpublished dissertation, University of London.

Storkey, Marian. 2000. "Using the schools' language data to estimate the total number of speakers of London's top 40 languages," in *Multilingual Capital: The Languages of London's Schoolchildren and Their Relevance to Economic, Social and Educational Policies*, ed. Philip Baker and John Eversley, 5–60. London: Battlebridge Publications.

Sundkler, Bengt. [1948] 1961. *Bantu Prophets in South Africa*. London: Oxford University Press.

Tardits, Claude. 1996. "Pursue to attain: a royal religion," in *African Crossroads*, ed. Ian Fowler and David Zeitlyn, 141–164. Oxford: Berghahn Books.

Taylor, Charles. 1994. "The politics of recognition," in *Multiculturalism: Examining the Politics of Recognition*, ed. Charles Taylor et al., 25–73. Princeton, NJ: Princeton University Press.

Tempels, Placide. [1945] 1959. *Bantu Philosophy*. Paris: Présence Africaine.

Ter Haar, Gerrie. 1998. *Halfway to Paradise: African Christians in Europe*. Cardiff: Cardiff Academic Press.

Teriba, A. and O.A. Phillips. 1971. "Income distribution and national integration." *Nigerian Journal of Economic and Social Studies* 13:1, 77–122.

Thomas, Keith. 1971. *Religion and the Decline of Magic*. London: Weidenfeld & Nicholson.

Thompson, Robert Farris. 1970. "The sign of the divine king." *African Arts* 3:3, 8–17 and 74–80.

Toulis, Nicole Rodriguez. 1997. *Believing Identity: Pentecostalism and the Mediation of Jamaican Ethnicity and Gender in England*. Oxford: Berg.

Turner, Edith. 1987. *The Spirit and the Drum*. Tucson: The University of Arizona Press.

Turner, Harold. W. 1965. *Modern African Religious Movements: An Introduction for the Christian Churches*. Nsukka: University of Nigeria.

―― 1967. *African Independent Church* (2 vols.). Oxford: Clarendon Press.

―― 1979. *Religious Innovation in Africa: Collected Essays on New Religious Movements*. Boston: G.K. Hall & Co.

Turner, Victor. 1967. *The Forest of Symbols: Aspects of Ndembu Ritual*. Ithaca: Cornell University Press.

―― 1975. *Revelation and Divination in Ndembu Ritual*. Ithaca: Cornell University Press.

Turner, Victor and Edward Bruner (eds.). 1986. *The Anthropology of Experience*. Chicago: University of Illinois Press.

Tutuola, Amos. [1948] 1989. *The Wild Hunter in the Bush of Ghosts*. Washington: Three Continents Press.

United Kingdom Council for Overseas Student Affairs (UKCOSA). 1972. *The Situation of Married Overseas Students in the U.K.* London: UKCOSA.

Valla, Jean-Pierre and Raymond H. Prince. 1989. "Religious experiences as self-healing mechanisms," in *Altered States of Consciousness and Mental Health*, ed. Colleen A. Ward, 149–166. London: Sage Publications.

Verger, Pierre. 1966. "The Yoruba High God—a review of the sources." *Odu* 2:2, 19–40.

―― 1969. "Trance and convention in Nago-Yoruba spirit mediumship," in *Spirit Mediumship and Society in Africa*, ed. John Beattie and John Middleton, 50–66. London: Routledge & Kegan Paul.

Wallman, Sandra. 1978. "The boundaries of 'race': processes of ethnicity in England." *Man* 13:2, 200–217.

Ward, Colleen A., 1989. "The cross-cultural study of altered states of consciousness and mental health," in *Altered States of Consciousness and Mental Health*, ed. Colleen Ward, 15–35. London: Sage.

Wavell, Stewart, Audrey Butt, and Nina Epton. 1966. *Trances*. London: George Allen & Unwin.

Weber, Max. [1922] 1966. *The Sociology of Religion*. London: Methuen.

Webster, Hutton. 1948. *Magic: A Sociological Study*. Stanford, CA: Stanford University Press.

Weinreich, P. 1986. "The operationalisation of identity theory in racial and ethnic relations," in *Theories of Race and Ethnic Relations*, ed. John Rex and David Mason, 299–320. Cambridge: Cambridge University Press.

Wescott, Joan and Peter Morton-Williams. 1962. "The symbolism and ritual context of the Yoruba Laba Shango." *Journal of the Royal Anthropological Institute* 92, 23–37.

Whyte, Susan Reynolds. 1991. "Knowledge and power in Nyole divination," in *African Divination Systems: Ways of Knowing*, ed. Philip Peek, 153–172. Bloomington: Indiana University Press.

Wilhelm, Richard (trans.). 1968. *The I Ching or Book of Changes*. London: Routledge & Kegan Paul Ltd.

Williams, Gavin. 1970. "The social stratification of a neo-colonial economy in Western Nigeria," in *African Perspectives*, ed. C. Allen and R. Johnson, 225–250. Cambridge: Cambridge University Press.

———— 1976. "Nigeria: a political economy," in *Nigeria: Economy and Society*, ed. Gavin Williams. London: Rex Collings.

Willis, Roy G. 1968. "Changes in mystical concepts and practices among the Fipa." *Ethnology* 7, 139–157.

Wilson, Bryan R. 1961. *Sects and Society: The Sociology of Three Religious Groups in Britain*. London: Heinemann.

———— 1973. *Magic and the Millennium: A Sociological Study of Religious Movements of Protest among Tribal and Third-World Peoples*. London: Heinemann.

———— 1985. "Secularization: the inherited model," in *The Sacred in a Secular Age*, ed. Phillip Hammond, 9–20. Berkeley: University of California Press.

———— 1990. *The Social Dimensions of Sectarianism: Sects and New Religious Movements in Contemporary Society*. Oxford: Clarendon Press.

Worsley, Peter. 1957. *The Trumpet Shall Sound*. London: McGibbon & Kee.

Zeitlyn, David. 1987. "Mambila divination." *Cambridge Anthropology* 12:1, 21–51.

———— 1990. "Professor Garfinkel visits the soothsayers: enthnomethodolgy and Mambila divination." *Man* 25:4, 654–666.

Aladura Literature

Abidoye, S. n.d. [1971]. *The Principles of the Holy Order of Cherubim and Seraphim Church*. London.

———— n.d. [1993, rev. ed.]. *Ready Made Prayer Book*. London.

———— n.d. [2001]. *Cherubim and Seraphim Church in Mellinnium Year*. London: Sawdexn Enterprises.

Abiodun Emanuel, Mrs C. *Celestial Vision of Her Most Rev. Mother Capt. Mrs C. Abiodun Emanuel, which originated Cherubim and Seraphim in 1925*. Yaba: Charity Press.

Akindele, J.M. n.d. *The Use and Efficacy of the Psalms*. Lagos: Alebiosu Press.

Chirwa, W.M. 1990. *The Establishment of Cherubim and Seraphim Church in the United Kingdom and Overseas*. London: C&S Church (UK).

Cherubim and Seraphim Church (UK). n.d. *General Codes of Conduct.* Mimeo.

Cherubim and Seraphim Council of Churches (UK). 1990. *Programme for the Silver Jubilee Celebrations.*

Cherubim and Seraphim, Eternal Sacred Order of Morning Star and St Michael Star, Fountain of Life, Mount Zion. n.d. *Holy Hymn Book.* Suru-lere, Lagos: Glory of the Morning Star Press.

Cherubim and Seraphim Movement. n.d. *Ofin, Ilana at Eto Ẹgbẹ Mimọ Kerubu ati Serafu* (Rules and Regulations of the Holy Order of Cherubim and Seraphim). Kaduna: Cherubim and Seraphim Press.

Cherubim and Seraphim Church Movement. 2000. *Europe District Year Book.*

Cherubim and Seraphim Movement of Northern Nigeria. ca.1965. *The Constitution, Principles and Regulations.* Kaduna: Baraka Press.

——— [1964 and 1965]. 24th and 25th Annual Conference. Kaduna: Cherubim and Seraphim Press.

——— 1967. *Daily Bible Reading Verses for Cherubim and Seraphim Movement of Northern Nigeria for 1967.* Kaduna: Cherubim and Seraphim Press.

Ẹgbẹ Mimọ Kerubu Ati Serafu Agbaiye. n.d. *Iwe Orin Mimọ* (Holy Hymn Book). Ogbomosho: Akande Printing Press.

Eternal Sacred Order of the Cherubim and Seraphim. 1930. *The Cherubim and Seraphim Memorandum and Articles of Association.* Nigeria: Ebute-Metta.

——— n.d. *History of Moses Orimolade at Ojokoro.* Ebute-Metta: Mount Zion (Religious) Enterprises. Lagos: Printed Hope Printing Press.

——— n.d. *The "Order," Rules and Regulations.* Lagos: Mount Zion (Religious) Enterprises.

——— n.d. *The "Order," Rules and Regulations, Duties of workers and Forty-Days Lenten Programme for "Cherubim and Seraphim."* Lagos: Mount Zion (Religious) Enterprises.

——— 1965. *Daily Bible Reading Pamphlet.* Ebute-Metta: Mount Zion (Religious) Enterprises.

Kalejaiye, S.O. n.d. [ca. 1964/65]. *Itan Ẹgbẹ K & S Kaduna.* Kaduna.

Khita, T.E.N. 1955. *Christianity and Its Hidden Facts.* Umuahia: National Printing Works.

Oshitelu, J.O. n.d. *The Book of Prayer with Uses and Power of Psalms and Precious Treasures Hidden Therein.* Ode Remo: Degosen Printing Works.

Phillips, H.A. 1962. *Ilana ati Eto Isin Kikun Ẹgbẹ Mimọ Kerubu ati Serafu. (Full Order of Service).* Ebute-Metta: T.U.P.P.

Sonde, J.O. n.d. *The Uses of Psalms, Efficacy of Prayers, and Mystery of Mysteries.* K&S, ESO Mount Zion. Ebute-Metta: Asalu Press.

Index

Printed in the United States
By Bookmasters